A LEVEL
Questions and Answers

D1324693

PHYSICS

Graham Booth

SERIES EDITOR: BOB McDUELL

Contents

HOW TO USE THIS BOOK

The aim of the *Questions and Answers* series is to provide the student with the help required to attain the highest level of achievement in important examinations. This book is intended to help you with A- and AS-level Physics or, in Scotland, Higher Grade Physics. The series is designed to help all students up to A grade. It relies on the idea that an experienced Examiner can provide, through examination questions, sample answers and advice, the help a student needs to secure success. Many revision aids concentrate on providing factual information which might have to be recalled in an examination. This series, while giving factual information in an easy-to-remember form, concentrates on the other skills which need to be developed for A-level examinations.

The *Questions and Answers* series is designed to provide:

- Easy-to-use **Revision Summaries** which identify important factual information. These are to remind you, in summary form, of the topics you will need to have revised in order to answer exam questions.
- Advice on the different types of question in each subject and how to answer them well to obtain the highest marks.
- Information about other skills, apart from the recall of knowledge, that will be tested on examination papers. These are sometimes called **Assessment Objectives**. Modern A-level examinations put great emphasis on the testing of other objectives apart from knowledge and understanding. Typically, questions testing these Assessment Objectives can make up 50% of the mark allocated to the written papers. Assessment Objectives include communication, problem solving, data handling, evaluation and interpretation. The *Questions and Answers* Series is intended to develop these skills by the use of questions and showing how marks are allocated.
- Many examples of **examination questions**, with spaces for you to fill in your answers. Students can improve their results by studying a sufficiently wide range of questions, providing they are shown the way to improve their answers to these questions. It is advisable that students try these questions first before going to the answers and the advice which accompanies the answers. Some of the questions come from actual examination papers or specimen materials issued by Examination Boards. Other questions have been written to closely mirror the types of question written for Examination Boards. The questions meet the requirements of all British Examination Boards.
- **Sample answers and mark schemes** for all of the questions.
- **Advice from Examiners**. By using the experience of Examiners we are able to give advice which can enable students to see how their answers can be improved and success be ensured.

Success in A-level examinations comes from proper preparation and a positive attitude to the examination developed through a sound knowledge of facts and an understanding of principles. The books are intended to overcome 'examination nerves' which often come from a fear of not being properly prepared.

THE IMPORTANCE OF USING QUESTIONS FOR REVISION

Past examination questions play an important part in revising for examinations. However, it is important not to start practising questions too early. Nothing can be more disheartening than trying to do a question which you do not understand because you have not mastered the topic. Therefore, it is important to have studied a topic thoroughly before attempting any questions on it.

How can past examination questions provide a way of preparing for the examination? First, it is unlikely that any question will appear in exactly the same form on the papers you will be sitting. However, there are restrictions to what can be set, as questions must cover the whole syllabus and test certain Assessment Objectives. The number of totally original questions that can be set on any part of the syllabus is therfore very limited and so similar ideas recur. It certainly will help you if the question you are trying to answer in an examination is familiar and you know you have done similar questions before. This is a great boost for your confidence, and this is what is required for examination success.

Secondly, practising examination questions will also highlight gaps in your knowledge and understanding which you can then go back and revise more thoroughly. It will indicate which sorts of questions you can do well and which, if there is a choice, you should avoid.

Thirdly, attempting past questions will get you used to the type of language used in questions.

Finally, having access to answers (as in this book) will enable you to see clearly what is required by the examiner, how best to answer each question, and the amount of detail required. Attention to detail is a key aspect of achieving success at A-level.

MAXIMISING YOUR MARKS

One of the keys to examination success is to know how marks are gained and lost by candidates. There are two important aspects to this: ensuring you follow the instructions on the examination paper and understanding how papers are marked.

Often candidates fail to gain the marks they deserve because they fail to follow the instructions. For instance, if you are told to answer four questions from a section and you answer five, you can receive credit for only four. The examiner may be instructed to mark the first four only and cross out additional ones. It would be unfortunate if the fifth answer was your best. Attempting too many questions also means you will have wasted time; you cannot have spent sufficient time on each of the other four questions and your answers to these could have suffered as a result.

Where a choice of questions is possible, candidates often choose the wrong ones. A question which looks familiar may not always be as easy as it seems and valuable time can be lost going up 'dead ends'. If you have a choice, spend time, before you begin writing, reading all of the questions and making rough notes. Then start with the questions you think you can do best and leave any of which you are unsure until later. In choosing, look at the marks allocated to various parts of the questions and try to judge whether you are confident in those parts which gain the most marks.

For every examination paper there is a 'mark scheme' which tells the examiner where marks should and should not be awarded. For example, where a question is worth a maximum of five marks, there will be five, six or maybe more points which gain marks and the examiner will award a mark for each of these given by a candidate, up to a maximum of five. Therefore a '(5)' shown after a question on an exam paper is an indication that the answer should contain five points. Obviously, lengthy writing will not gain credit unless the candidate includes the correct points. Try therefore to keep your answers brief and to the point. Look at your answers critically after you have written them and try to decide how many different important points you have made.

An important principle of examination marking is called 'consequential marking'. This means that, if a candidate makes a mistake, the examiner must penalise the candidate for it only once. For example, if you made a mistake early in a calculation so that you came up with an incorrect value, you would obviously lose a mark. However, if you then used this incorrect value in a later part of the question, and your working was correct apart from this incorrect value, you would not lose any more marks. Therefore, always write down all of your workings, so that you can gain marks even if you make an early slip-up. You will see examples of consequential marking in the questions in this book.

As a rough guide, in an examination you should be aiming to score a mark each minute. When about 15 minutes of the examination remain, it is worth checking to see whether you are short of time. If you feel you are seriously running out of time, it is very important to try to score as many marks as possible in the time that remains. In any question worth five marks, one or two will be easily scored and one or two will be very difficult to score. Concentrate on scoring the easy marks on each of the questions that remain. Do not try to write sentences at this point, just put the main points down clearly in note form.

DIFFERENT TYPES OF EXAMINATION QUESTION

Multiple-choice questions

These questions require you to make a choice from a number of answers, usually four or five. It is important that you examine all the answers before making your choice. Do not be tempted to choose the answer that 'looks right', as this is often a distractor, intended to appeal to the candidate who does not have the depth of knowledge and understanding required to answer the question. Although not all examination boards use multiple choice questions, examples have been included in this book because they convey important physical principles as well as providing useful revision practice.

Structured questions

These are the most common type of question used in A-level Physics papers. Most of the questions in this book are structured questions. The reason why they are so widely used is that they are versatile. They can be short with little opportunity for extended writing. In this form they are suitable for testing knowledge, understanding and the use of physical relationships. Alternatively, they can be longer and more complex in their structure, with opportunities for extended writing and demonstration of higher level skills of interpretation and evaluation, as well as testing the application of knowledge in familiar and unfamiliar contexts.

In a structured question, the question is divided into parts (a), (b), (c) etc. These parts can be further subdivided into (i), (ii), (iii), (iv) etc. A structure is built into the question and hence into your answer. This is where the term structured question comes from.

For each part of the question there are a number of lines or a space for your answer. This is a guide to you about the detail required in the answer but it does not have to limit you. If you require more space continue your answer on a separate sheet of paper but make sure you label your answer clearly, e.g. 3(a)(ii).

For each part of the question there is a number in brackets, e.g. (3), to show you how many marks are allocated to this part of the question by the examiner. If a part is worth three marks, for example, the question requires more than one or two words. As a general rule, if there are three marks allocated, you will need to make three points.

To give you a guide as you work through structured questions, papers are often designed to enable you to score one mark per minute. A question worth a maximum of fifteen marks should therefore take about fifteen minutes to answer.

You do not have to write your answers in full sentences. Concise notes are often the most suitable response; the important thing is that you communicate your ideas clearly and effectively to the examiner. Where a part of a question is worth three or more marks it is a good idea to plan your solution or description in a logical order. This is particularly important when describing experimental work.

It is most important to read the stimulus material in the question thoroughly and more than once. This information is often not used fully by candidates and, as a result, the question is not answered fully. The key to answering many of these questions comes from the appreciation of the

full meaning of the 'command word' at the start of the question – 'state, describe, explain'. The following glossary of command words may help you in the answering of structured questions.

- **State** This means a brief answer is required, with no supporting evidence. Alternatives include **write down, give, list, name**.
- **Define** Just a definition is required.
- **State and explain** A short answer is required (see **state**), but then an explanation is required. A question of this type should be worth more than one mark.
- **Describe** This is often used with reference to a particular experiment. The important points should be given about each stage. Again this type of question is worth more than one mark.
- **Outline** The answer should be brief and the main points picked out.
- **Predict** A brief answer is required, without supporting evidence. You are required to make logical links between various pieces of information.
- **Complete** You are required to add information to a diagram, sentence, flow chart, graph, key, table, etc.
- **Find** This is a general term which may mean calculate, measure, determine, etc.
- **Calculate** A numerical answer is required. You should show your working in order to get an answer. Do not forget the correct units.
- **Suggest** There is not just one correct answer, or you are applying your answer to a situation outside the syllabus.

Free response questions

These can include essay questions. In this type of question you are given a question which enables you to develop your answer in different ways. Your answer is strictly free response and you write as much as you wish. Candidates often do not write enough or try to 'pad out' the answer. Remember you can only score marks when your answer matches the marking points on the examiner's marking scheme.

In this type of question it is important to plan your answer before starting it, allocating the correct amount of time to each part of the question. Again, you should present your ideas in a logical order.

Sample question

Describe a laboratory experiment that you could carry out to determine the value of g, free-fall acceleration. Explain how you would calculate the value of g from the readings that you would take. (10)

Plan

First choose a suitable method. This could be by using a simple pendulum or by timing a freely-falling object.
The description needs to start by making clear the method being used.
Next there should be a statement of what is to be varied and how this is to be done. This is followed by a list of the measurements or readings to be taken, together with any necessary detail.
At this level it is important to describe precautions to take to ensure reliability of the result, so it is necessary to include details of any precautions taken to reduce the effects of experimental error.
Finally, explain how the final value is calculated.

Here is a sample answer.
Free-fall acceleration, g, can be measured by timing a falling object over a measured distance.
At this stage put in a diagram to illustrate the chosen method. Do not then repeat in words the details that are given on the diagram.

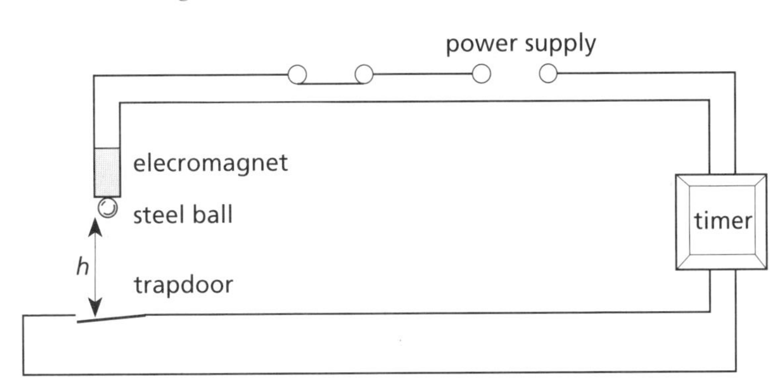

When the electromagnet is switched off the ball is released and the timer starts to time. The timer stops timing when the ball causes the trapdoor to open.
One mark is awarded here for a valid method plus one mark for realising that the time needs to be measured and a further mark for details of how this is done, making three marks altogether so far.
The height, h, is measured using a ruler and the experiment is repeated twice, the time being recorded for each drop of the ball.
Two marks are awarded here, one for stating that the height is measured and a second for details of how this is done.
The electromagnet is raised to change h so that several sets of results are taken over a range of values of h from 0.5 m to 1.5 m.
One mark for stating the quantity to be varied and one mark for details of how to do this.
For each value of h an average time, t, is calculated and a graph is drawn of h against $\frac{1}{2}t^2$. The gradient of the graph represents g, free-fall acceleration.
One mark for having repeated the experiment at the same value of h and averaged the times. There are two marks for arriving at the result; one is for stating what is to be plotted and the second is for details of how to arrive at the final value.

ASSESSMENT OBJECTIVES IN PHYSICS

Assessment Objectives are the intellectual and experimental skills you should be able to show. Opportunities must be made by the Examiner when setting the examination paper for you to demonstrate your mastery of these skills when you answer the question paper.

Knowledge and understanding can only contribute about half of the marks on the written paper. The other half of the marks are acquired by mastery of the other Assessment Objectives described below.

1. Communicate scientific observations, ideas and arguments coherently and effectively (*Weighting on papers approximately 5 – 10%*)

In any examination, questions are built into the paper to test your ability to communicate scientific information. Often these questions require extended answers.

In this type of question it is important to look at your answer objectively after you have written it and try to judge whether your answer is communicating information effectively.

2. Select and use information to identify patterns and trends and translate data from one form to another (*Weighting on papers approximately 10 – 15%*)

In questions testing this Assessment Objective you are often asked to pick information from a chart or table and use it in another form, e.g. to draw a graph. You may be asked to deduce a physical relationship using information from a graph.

It is important to transfer the skills you have acquired in Mathematics to your work in Physics. In particular, for graphical work, you need to be able to:

- Select and plot two variables from data.
- Choose appropriate scales for graph plotting.
- Choose and draw the best straight line or smooth curve on a graph.
- Determine the gradient and intercept of a linear graph and relate them to physical quantities.
- Understand and use $y = mx + c$.
- Understand the possible physical significance of the area below a curve and be able to calculate it or measure it by counting squares as appropriate.
- Use logarithmic plots to test exponential and power law variations.

3. Interpret, evaluate and make informed judgements from relevant facts, observations and phenomena (*Weighting on papers approximately 10 – 15%*)

It is much easier to test this Objective on long questions than on short ones. The command word 'suggest' is very frequently used as the information given, perhaps in a paragraph or a table or a diagram or any combination of these, is open to more than one interpretation.

Look carefully at all of the information given and look for possible alternative interpretations before writing your answer.

4. Solve qualitative and quantitative problems (*Weighting on papers approximately 10 – 15%*)

Qualitative problems can include describing phenomena in terms of fundamental physical principles, using for example the laws of electromagnetic induction or Newton's laws of motion. When answering these questions it is important to state the principles being used and show clearly how they apply to the phenomenon that you are explaining.

Quantitative problems include the full range of physical calculations which have baffled students studying physics for generations. When attempting to carry out a physical calculation, remember:

❶ Show all of your working, so credit can be given if you do not get the correct answer, but get some way through the question.

❷ Take care when substituting in a formula to be consistent with your units.

❸ Give correct units to your answers if there are units. Remember ratios, including efficiencies, do not have units.

Formulae you should know

This is a list of formulae that you may need to use in answering Physics questions but will **not** be given to you either on the examination paper or on a separate formula sheet.

pressure = force ÷ area $p = \frac{F}{A}$

speed = distance ÷ time taken $v = \frac{s}{t}$

$\frac{\text{pressure}_1 \times \text{volume}_1}{\text{temperature}_1} = \frac{\text{pressure}_2 \times \text{volume}_2}{\text{temperature}_2}$ $\frac{p_1V_1}{T_1} = \frac{p_2V_2}{T_2}$

work done = force × distance moved in its own direction $W = F \times s$

power = energy transferred or work done ÷ time taken $P = \frac{E}{t}$ or $\frac{W}{t}$

moment of a force (or torque) = size of force × perpendicular distance from force to pivot

in a balanced system:
the sum of clockwise moments about a point = the sum of the anticlockwise moments about that point

energy = potential difference × current × time $E = IVt$

force = mass × acceleration $F = ma$

acceleration = increase in velocity ÷ time taken $a = \frac{v - u}{t}$

wave speed = frequency × wavelength $v = f\lambda$

charge = current × time $Q = It$

potential difference = current × resistance $V = IR$

electrical power = current × potential difference $P = IV$

weight = mass × gravitational field strength $W = mg$

kinetic energy = $^1/_2$ × mass × (speed)^2 ke = $^1/_2 mv^2$

change in gravitational potential energy = mass × gravitational field strength × change in height
gpe = $mg\Delta h$

momentum = mass × velocity $p = mv$

centripetal force = mass × speed^2 ÷ radius $F = \frac{mv^2}{r}$

potential difference = energy transferred ÷ charge $V = \frac{W}{Q}$

resistance = $\frac{\text{resistivity} \times \text{length}}{\text{cross sectional area}}$ $R = \frac{\rho l}{A}$

capacitance = $\frac{\text{charge stored}}{\text{potential difference}}$ $C = \frac{Q}{V}$

pressure × volume = number of moles × molar gas constant × absolute temperature $pV = nRT$

the inverse square laws for force and field strength in radial electric and gravitational fields:

$$F_E = \frac{1}{4\pi\varepsilon_0}\frac{Q_1Q_2}{r^2} \qquad E = \frac{Q}{4\pi\varepsilon_0 r^2} \qquad F_G = \frac{Gm_1m_2}{r^2} \qquad g = \frac{GM}{r^2}$$

1 Forces at rest

REVISION SUMMARY

Physical quantities such as mass, length, electric current, momentum and force can be divided into two groups; **vectors** and **scalars.** A scalar quantity is one that is described by its size only, for example the mass of a bag of sugar is 1 kg. A vector quantity is described by its size and direction, for example the velocity of the wind is 5 $m\,s^{-1}$ from the south.

Scalars can be added and subtracted using ordinary rules of number but it is not correct to write '2 N + 2 N = 4 N'. The result of adding together two 2 N forces can be anything between 0 N and 4 N, depending on the directions. The diagram illustrates the method of finding the sum of two vectors and the difference between two vectors.

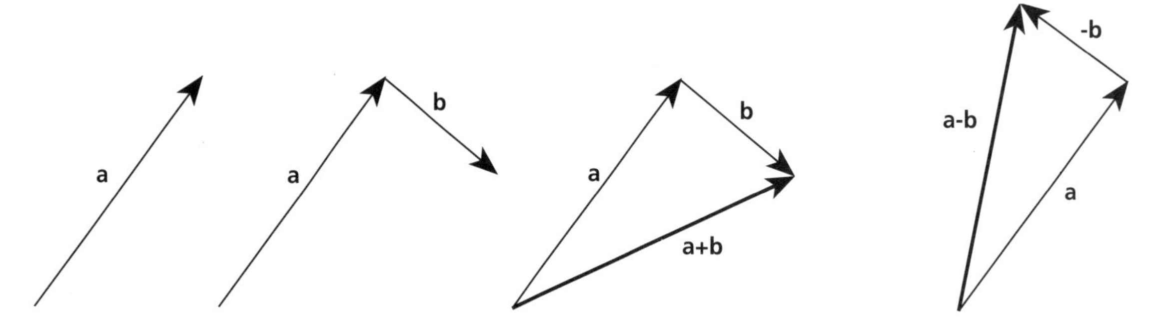

Any number of vectors can be added using this method; if the vectors form a closed polygon when they are drawn end-to-end then their sum is zero.

Vector quantities such as forces can have effects in any direction except for one which is at right angles to their own. The effective size of a vector in another direction is equal to the size of the vector multiplied by the cosine of the angle between the two directions. The diagram shows how this can be used to split, or **resolve,** a vector into two perpendicular components. The vector has an effect in each of these directions.

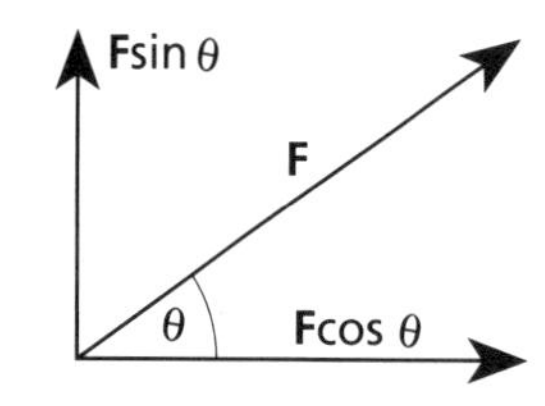

If you need to revise this subject more thoroughly, see the relevant topics in the Letts A level *Physics Study Guide.*

Forces can have a **turning effect** as well as a pushing or pulling effect. The turning effect, or **moment**, of a force is calculated using the formula:

moment of force = size of force × perpendicular distance from force to pivot

Under a system of coplanar forces, if an object is in **equilibrium** then the sum of the forces in any two mutually perpendicular directions within a plane must be zero and the sum of the clockwise moments about any pivot must equal the sum of the anticlockwise moments about the pivot. The last statement is known as the **principle of moments**.

The **gravitational force** acting on an object, or the **weight** of the object, is taken as acting from the **centre of gravity** or **centre of mass**. If the mass is uniformly distributed the centre of mass is at the centre of the object; uneven distribution shifts the centre of mass towards the more massive part of the object.

Pressure is defined as the perpendicular force acting per unit area,

$$\text{pressure} = \frac{\text{force}}{\text{area}}$$

Unlike solids, fluids exert pressure in all directions. The pressure due to a column of fluid is equal to height × density × gravitational field strength, or $h\rho g$.

QUESTIONS

Throughout this section take the value of g, free-fall acceleration, to equal 10 $m\,s^{-2}$.

1 Which pair of physical quantities consists of two **vectors**?

A speed and acceleration
B power and momentum
C mass and kinetic energy
D force and displacement

(1)

2 Object O is held in equilibrium by three coplanar forces **X**, **Y** and **Z** as shown. The force **X** makes an angle θ with the vertical.

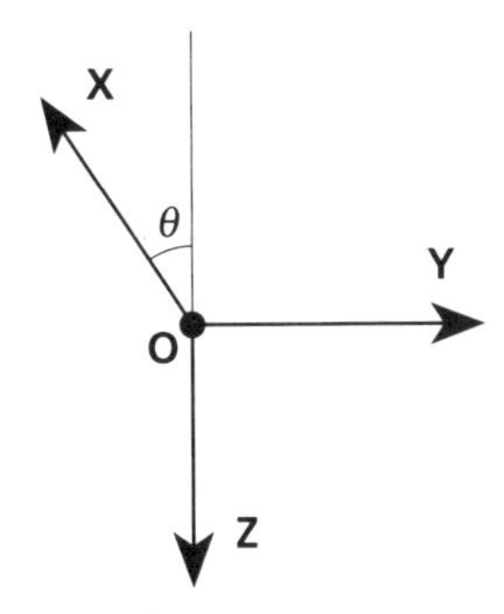

The magnitude of **Y** must be

A X−Z
B X tanθ
C X sinθ
D X cosθ

(1)
AEB

3 A cyclist free-wheels down a slope, inclined at 15° to the horizontal, at a constant velocity of 3 $m\,s^{-1}$.

The combined mass of the rider and bicycle is 70 kg. The total force of friction is

A 181 N
B 210 N
C 362 N
D 391 N
E 676 N

(1)
SEB

QUESTIONS

4 The diagram shows a diving-board held in position by two rods X and Y.

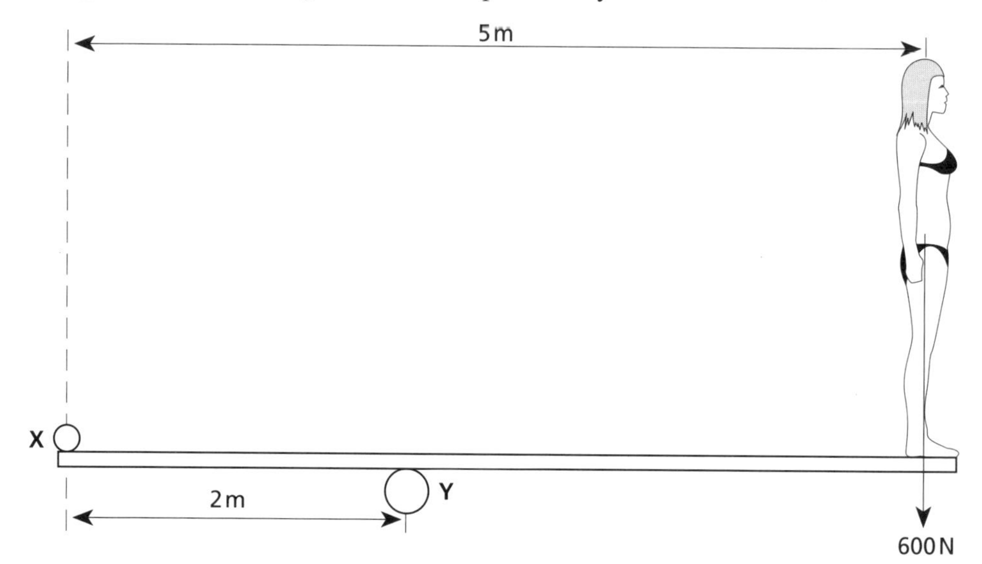

What additional forces do these rods exert on the board when a diver of weight 600 N stands on the right-hand end?

	At X (downwards)	At Y (upwards)
A	400 N	1000 N
B	600 N	1200 N
C	900 N	600 N
D	900 N	1500 N

(1)
Oxford

5 A tractor pulls a log at a steady speed in a straight line across level ground.

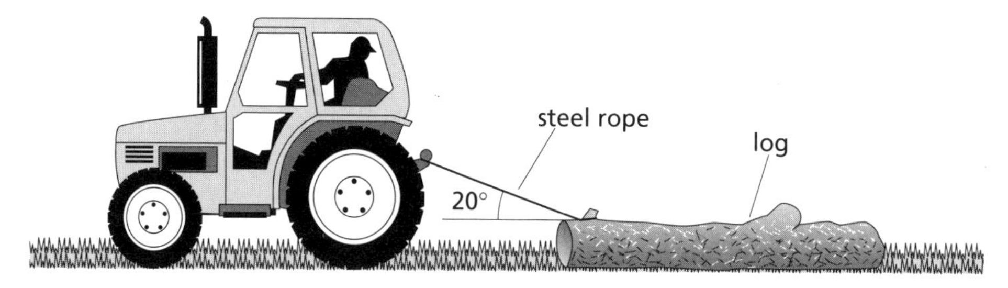

The steel rope between the tractor and the log makes an angle of 20° with the ground as shown. If the horizontal frictional force between the log and the ground is 500 N, calculate the tension in the steel rope.

....

....

(2)
SEB

QUESTIONS

6 In the leisure pursuit called parascending a person attached to a parachute is towed over the sea by a tow-rope attached to a motor boat, as shown in the left hand diagram.

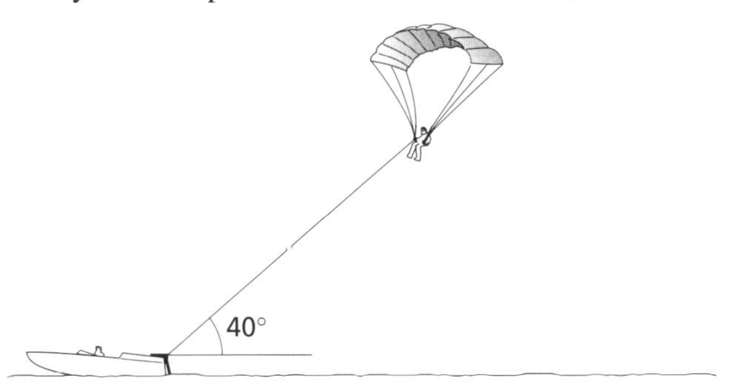

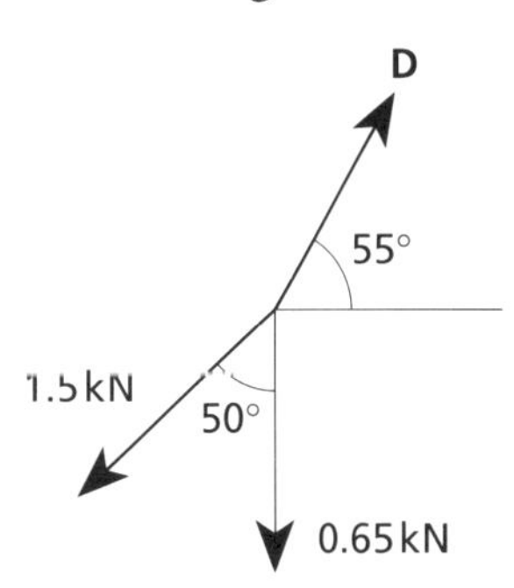

The right hand diagram shows the directions of the forces acting on a person of weight 0.65 kN when being towed horizontally at a constant speed of 8.5 $m\,s^{-1}$. The 1.5 kN force is the tension in the tow-rope and the force labelled D is the drag force.

(a) State why the resultant force on the person must be zero.

....

(b) Using a vector diagram, or otherwise, determine the magnitude of the drag force.

....

(c) (i) State the magnitude and direction of the force exerted **by** the tow-rope **on** the boat.

....

....

(ii) Determine the horizontal resistance to motion of the boat produced by the tow-rope.

....

(iii) The horizontal resistance to motion produced by the water is 1200 N. Determine the useful power developed by the boat's motor.

.... (8)

AEB

QUESTIONS

7 (a) State the difference between a *vector* and a *scalar* quantity.

....

....

(b) A device for removing tightly fitting screw tops from jars and bottles is shown in the diagram.

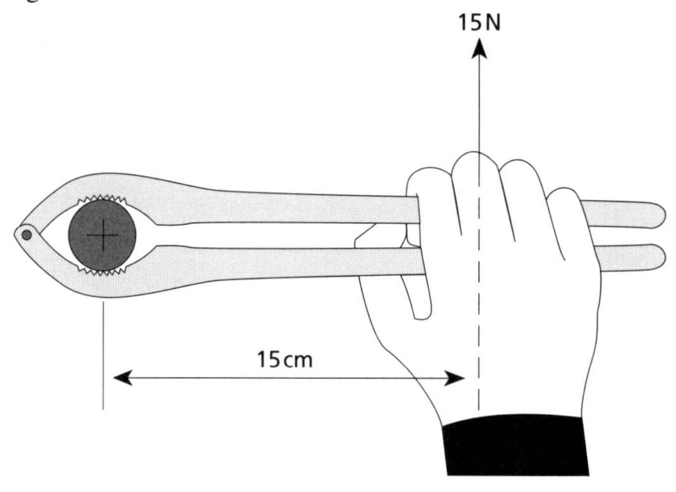

In one case a constant force of magnitude 15 N has to be applied.
The force is applied for one complete turn to remove the top.

(i) Calculate the torque (turning effect) which has to be applied to remove the top.

....

(ii) Calculate the work done when opening the bottle.

....

(c) A tight-fitting cork can be removed by pumping air into the bottle using the device shown below.

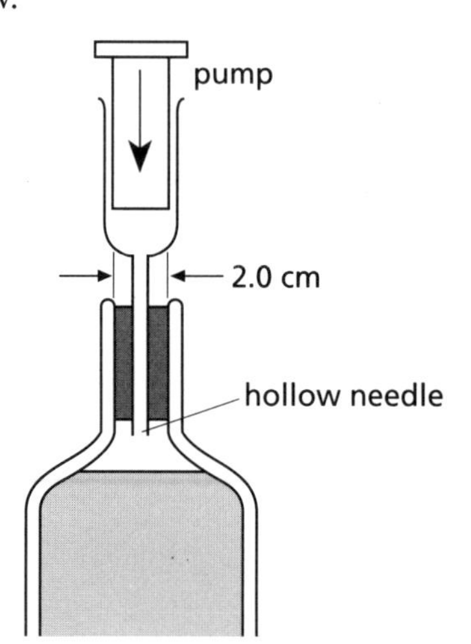

QUESTIONS

Assuming that the force which has to be overcome to remove the cork is 30 N, calculate the pressure inside the bottle when the cork begins to move.

Atmospheric pressure = 1.0×10^5 Pa.

.... (8)

AEB

8 (a) Physical quantities may be classified as scalars or as vectors. Tick (✓) the appropriate boxes in the table to indicate whether the listed quantities are scalars or vectors.

Quantity	Scalar	Vector
Displacement	☐	☐
Momentum	☐	☐
Energy	☐	☐
Magnetic flux density	☐	☐
Charge	☐	☐
Half-life	☐	☐

(3)

(b) A projectile is launched from a point on level ground with velocity v at an angle α above the horizontal.

(i) Sketch a labelled vector diagram to show the velocity of the projectile, the horizontal component of the velocity, and the vertical component of the velocity, at the moment of launching.

(3)

(ii) At what angle of projection α is the vertical height reached a maximum?

α = __________ ° (1)

(iii) At what angle of projection α is the horizontal range a maximum?

α = __________ ° (1)

(c) A stone is projected with a horizontal velocity of 40 m s^{-1} from the edge of a cliff 50 m high and strikes the ground at a point X, as shown in the diagram.

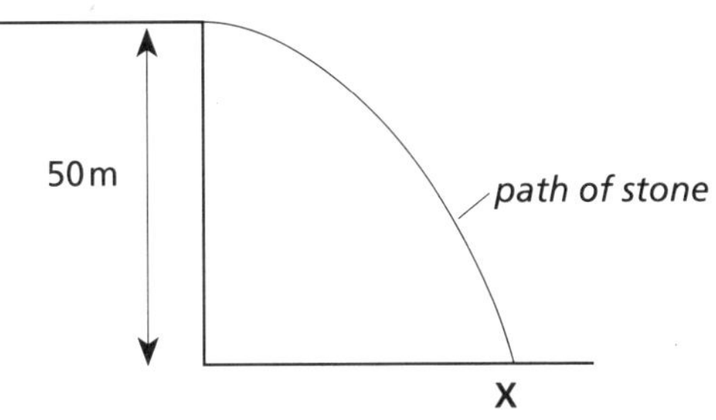

QUESTIONS

(i) In the space below *sketch* a vector diagram to show clearly the velocity of the stone, and the horizontal and vertical components of the velocity, *one second* after launch. Also, after necessary calculations, label the velocity vector and its components with their magnitudes.

(3)

(ii) Calculate the vertical component of the velocity of the stone just as it strikes the ground at X.

....

....

....

vertical component of velocity = __________ m s^{-1} (2)

(iii) In the space below draw a labelled diagram to scale to show the horizontal and vertical components of the velocity of the stone just as it strikes the ground at X.

(2)

(iv) Complete the construction and obtain, by measurement or calculation, the magnitude and direction of the stone's velocity as it strikes the ground at X.

Magnitude of velocity at X = __________ m s^{-1}

Direction of velocity at X = __________ ° to the horizontal (2)

NICCEA

Graphs can be used to represent the motion of objects. A **distance-time** graph shows how the total distance travelled changes with time. The gradient of a distance-time graph represents the speed of the object. **Displacement-time** graphs are different because displacement is a vector and can have both positive and negative values. The gradient of a displacement-time graph represents **velocity**, which is also a vector.

Speed-time and **velocity-time** graphs also differ in that velocity is a vector quantity. The gradient of a velocity-time graph represents the **acceleration** but that of a speed-time graph represents only the numerical value of the acceleration, not its direction. For both types of graph, the distance travelled is represented by the total area between the line and the time axis.

For objects moving with a constant (uniform) acceleration in a straight line, the following equations apply to the motion:

$$v = u + at \qquad v^2 = u^2 + 2as \qquad s = ut + \tfrac{1}{2}at^2$$

where a = acceleration, s = distance travelled, t = time, u = initial velocity and v = final velocity.

Some objects have motion in two directions which are independent of each other. A projectile (e.g. a ball thrown at any angle except vertically or a dart thrown at a dartboard) has an accelerated vertical motion, to which the above equations apply, but is moving at constant speed horizontally (assuming air resistance is neglected). The equation for the horizontal motion is distance travelled = speed × time, $s = vt$.

Air resistance and other resistive forces can usually be neglected when considering the motion of streamlined objects moving at low speeds. As the speed of an object increases, the resistive forces also increase; this causes both the resultant accelerating force and the acceleration to decrease. Terminal velocity, i.e. movement at a constant speed, is reached when the accelerating and resistive forces are balanced.

Forces which cause movement are doing **work**. The work done = force × distance moved in the direction of the force, $W = Fs$. This is also equal to the quantity of energy that is transferred when the force is working. The rate at which a force works or transfers energy is called the **power**. The equation for power is power = work done (or energy transfer)/time taken, $P = W/t$.

Gravitational potential energy (gpe for short) is energy due to position. Close to the Earth's surface, where there is very little change in the weight of an object at different heights above the surface, the change in gpe is $mg\Delta h$, where the symbol Δh means 'change in height' and g is free-fall acceleration.

Kinetic energy is energy due to motion. The formula for the kinetic energy of an object is $^1/_2mv^2$. Energy is **always** conserved, but in many processes it becomes spread out as thermal energy of the surroundings. The **efficiency** of a process measures what proportion of the energy ends up in a useful form; efficiency = useful work or energy output/ total work or energy input.

All moving objects have **momentum**. Momentum is a vector quantity which is calculated using the definition momentum = mass × velocity, $p = mv$. Momentum is a useful quantity because it is conserved (i.e. the total momentum remains the same) in collisions or interactions between two or more objects, provided that no external forces act to increase or decrease the momentum. This is known as the **principle of conservation of momentum**. Collisions where the total kinetic energy remains the same are called **elastic**. Collisions between gas molecules are elastic. In an inelastic collision, some of the kinetic energy is transferred into other energy forms.

Newton's first two **laws of motion** describe how the motion of an object depends on the resultant of the forces acting. The **first law** states that if an object is at rest or moving in a straight line at constant speed (i.e. constant velocity) then there is either no force acting on it or the forces are balanced.

REVISION SUMMARY

The **second law** states that the rate of change of momentum of an object is proportional to and in the same direction as the resultant of the forces acting. Two important results follow from the second law:

$$\text{force} = \text{mass} \times \text{acceleration} \qquad F = ma$$
$$\text{change in momentum} = \text{force} \times \text{time for which it acts} \qquad \Delta(mv) = Ft$$

The 'force × time for which it acts' is often called the **impulse** of a force and so is measured in N s.

The **third law** is the shortest and least well understood; it states that 'to every action there is an equal and opposite reaction'. This means that objects exert forces on each other, so that if 'A pulls B' then 'B pulls A with an equal size force in the opposite direction'.

An object moving at constant speed in a circle has a velocity which is continually changing and therefore it is accelerating. This acceleration is directed towards the centre of the circle and has a value of v^2/r, where v is the velocity and r is the radius of the circle. The unbalanced force needed to cause this acceleration is mv^2/r. This force is sometimes called the centripetal force but it is important to realise that it is not an extra force, but it comes from the imbalance of the forces acting. The 'centripetal' force when a bicycle goes round a corner, for example, comes from the friction force between the tyres and the road.

It is sometimes more convenient to use angular velocity, symbol ω, when dealing with circular motion. The **angular velocity** of an object does not depend on the radius of the circle. Angular velocity is measured in radians per second, where one revolution is equal to an angle of 2π radians. The relationship between linear velocity and angular velocity is $v = r\omega$ and the equations for circular motion become acceleration = $r\omega^2$ and resultant force = $mr\omega^2$.

If you need to revise this subject more thoroughly, see the relevant topics in the Letts A level *Physics Study Guide*.

Throughout this section take the value of g, free-fall acceleration, to equal 10 m s^{-2}.

QUESTIONS

1 A car travelling at 30 m s^{-1} starts to brake when it is 50 m from a stationary lorry. The car moves in a straight line and manages to stop just before reaching the lorry.
What is the deceleration of the car, in m s^{-2}?

A 0.6
B 4.5
C 9
D 10
E 18

(1)
SEB

2 The driver of a car has a mass of 80 kg. Whent the car brakes sharply the deceleration of the car and driver is 2.5 m s^{-2}. The size of the total force exerted by the car seat and safety belt on the driver during braking is:

A 200 N
B 600 N
C 825 N
D 1000 N

(1)

3 A force, which is applied in a straight line to an object, varies with time as shown in the following graph.

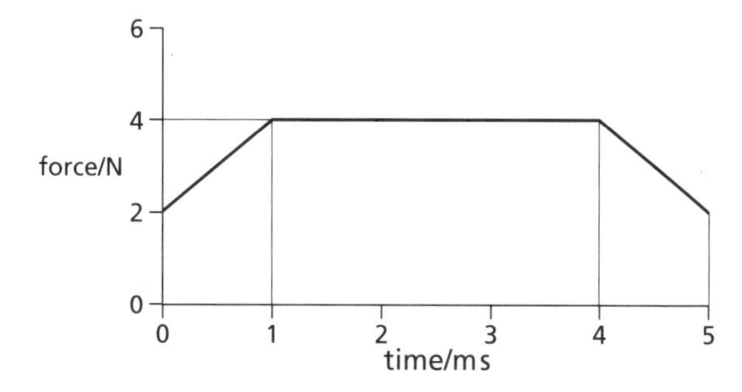

What is the total impulse given to the object by the force in this 5 millisecond time interval?

A 8×10^{-3} Ns
B 10×10^{-3} Ns
C 15×10^{-3} Ns
D 18×10^{-3} Ns
E 20×10^{-3} Ns

(1)
SEB

4 A ball of mass 0.2 kg is attached to a string and swung in a vertical circle of radius 0.5 m. What is the tension, in N, in the string when the ball passes through its lowest point at a speed of 5 m s^{-1}?

A 2
B 8
C 10
D 12

(1)
AEB

QUESTIONS

5 The engine of a car of mass 1200 kg works at a constant rate of 18 kW. The top speed of the car is 30 m s^{-1}.

(a) Find the resistance to its motion at the top speed.

....

....

....

(b) Assuming that the resistance to the motion is proportional to the speed of the car, calculate the acceleration of the car at the instant when its speed is 10 m s^{-1}.

....

....

....

....

....

.... (7)

WJEC

6 A tennis ball of mass 60 g leaves a player's racket horizontally at a speed of 48 m s^{-1}.

(a) (i) Assuming that the ball is at rest before it is struck by the racket, calculate the momentum given to the ball.

....

.... (2)

(ii) The ball is in contact with the racket for 8.0 ms. Calculate the mean value of the propulsive force experienced by the ball.

....

.... (2)

(iii) Calculate the mean acceleration of the ball during this period.

....

.... (1)

(b)

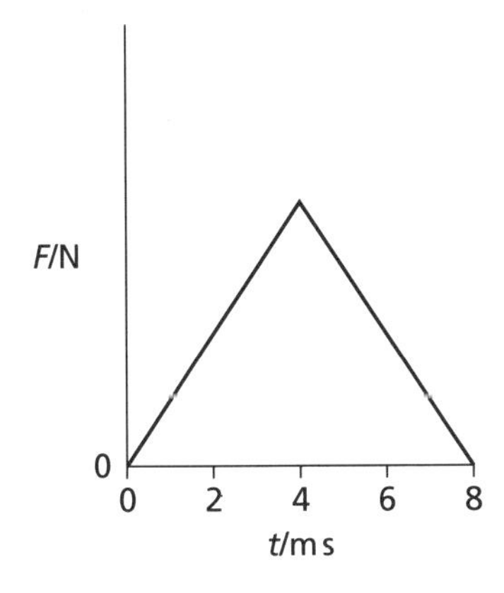

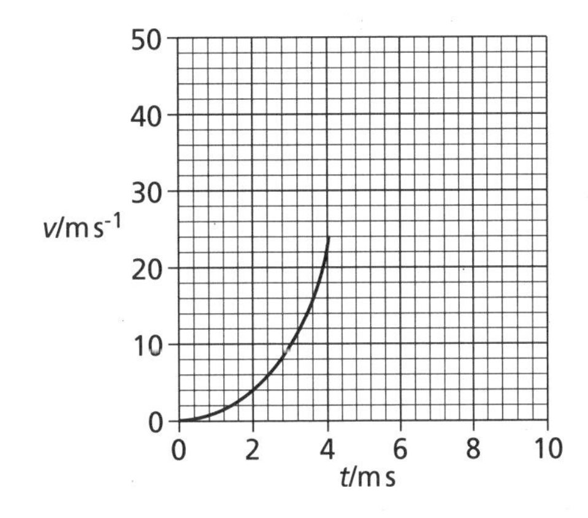

The sketch graph shows how the force on the ball varies while it is in contact with the racket. The graph shows how the speed, v, of the ball varies during the first 4 ms of its movement.

On the same diagram, continue the curve to show how the speed varies in the next 6 ms. Assume that resistances to motion are negligible. (3)

(c) Explain how you would use your completed graph to find the distance travelled by the ball while it is in contact with the racket.

....

.... (2)

Oxford

7 A model car of mass 99 g rests at the bottom of a slope as shown in the diagram.

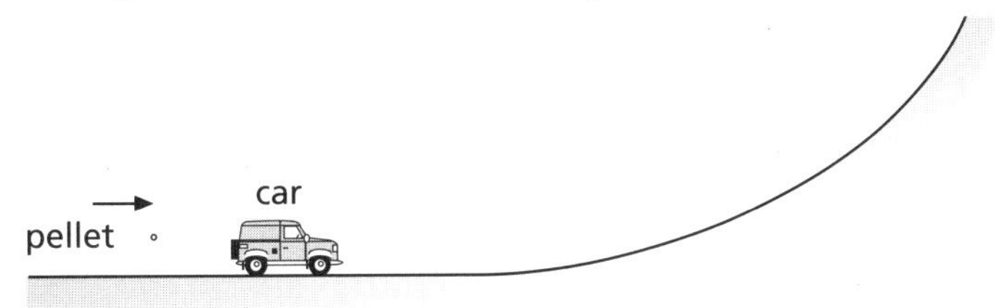

A pellet of mass 1 g is fired horizontally at a velocity of 200 m s^{-1} into the model car and remains embedded.

(a) Name the type of collision.

....

(b) To what vertical height above its initial position will the car rise, assuming it moves freely up the slope?

....

....

....

.... (5)

SEB

QUESTIONS

8 A lunar landing craft descends vertically towards the surface of the Moon with a constant speed of 2.0 m s^{-1}. The craft and crew have a total mass of 15 000 kg. Assume that the gravitational field strength on the Moon is 1.6 N kg^{-1}.

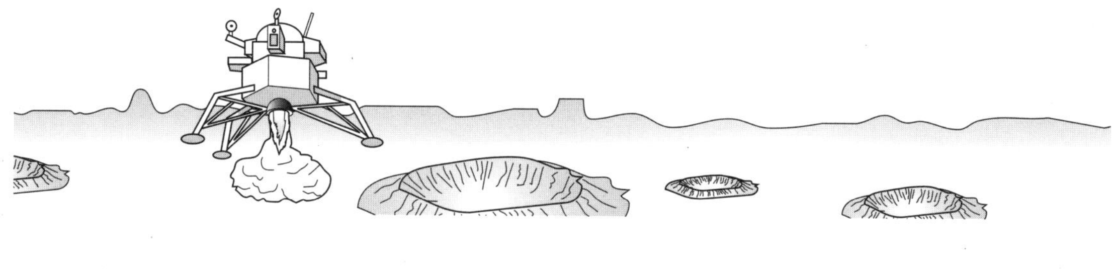

(a) During the first part of the descent the upward thrust of the rocket engines is 24 000 N. Show that this results in the craft moving with a constant speed.

....

....

.... (2)

(b) The upward thrust of the engine is increased to 25 500 N for the last 18 seconds of the descent.

(i) Calculate the deceleration of the craft during this time.

....

....

....

(ii) What is the speed of the craft just before it lands?

....

....

(iii) How far is the craft above the surface of the Moon when the engine thrust is increased to 25 500 N?

....

.... (7)

SEB

QUESTIONS

9 (a) State what is meant by *angular velocity*.

....

.... (2)

(b) A stone is tied to one end of a cord and then made to rotate in a horizontal circle about a point C with the cord horizontal, as shown in the diagram
The stone has speed v and angular velocity ω about C.

(i) Write down a relation between the speed v, the length r of the cord and the angular velocity ω.

....

(ii) Explain how v can be made to vary when ω is constant.

....

....

(iii) Explain why there needs to be a tension in the cord to maintain the circular motion.

....

....

(iv) Write down an expression for the acceleration of the stone in terms of v and r.

....

Hence, if the stone has mass m, show that the tension T in the cord is given by

$$T = mv\omega.$$

....

....

.... (8)

(c)

12 m s^{-1}

7.0 m

passenger enters loop →

passenger leaves loop →

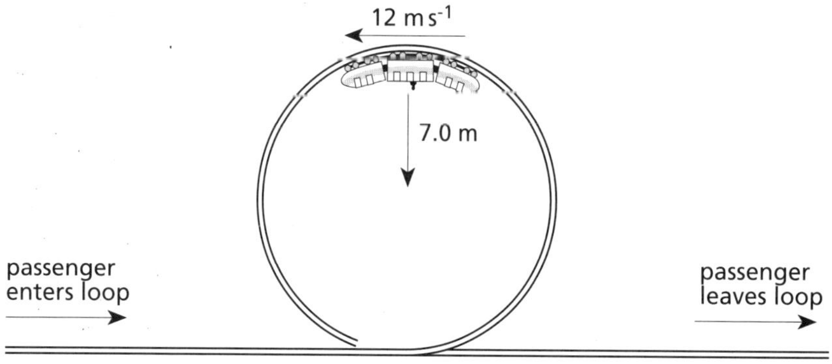

On one particular ride in an amusement park, passengers 'loop-the-loop' in a vertical circle, as illustrated in the diagram.
The loop has a radius of 7.0 m and a passenger, mass 60 kg, is travelling at 12 m s^{-1} when at the highest point of the loop. Assume that frictional forces may be neglected.

(i) Calculate, for the passenger when at the highest point,
 (1) the centripetal aceleration,
 (2) the force the seat exerts on the passenger.

....

....

....

(ii) The passenger now moves round and descends to the bottom of the loop. Calculate
 (1) the change in potential energy of the passenger in moving from the top of the loop to the bottom,
 (2) the speed of the passenger on leaving the loop.

....

....

....

....

....

(iii) Operators of this ride must ensure that the speed at which the passengers enter the loop is above a certain minimum value. Suggest a reason for this.

....

....

.... (10)

UCLES

REVISION SUMMARY

All objects can be made to vibrate by being in contact with another vibrating object; for example the casing of a washing machine vibrates at the frequency of rotation of the drum, whatever that frequency is. This is an example of a **forced** vibration. A **free** vibration occurs when an object is pushed or displaced and allowed to vibrate at its **natural frequency**; a child on a swing is an example of this.

Any vibrating object is subject to resistive forces that cause **damping** of the vibrations; the amplitude decreases as energy is transferred to the surroundings. **Resonance** occurs when the forcing vibration is at the same frequency as the natural frequency; this allows a large **amplitude** vibration to be maintained because the vibrating object is able to absorb energy to replace that lost through damping. An object is said to be **critically damped** if the resistive forces are just large enough to stop vibration following a displacement from its normal position. A motor vehicle suspension is an example of **critical damping**.

Simple harmonic motion (shm) is the simplest form of oscillation to analyse mathematically. The two conditions for a motion to be classed as simple harmonic are that the acceleration must be proportional to the displacement from a fixed point and must always be directed towards that point. A mass on a spring and a pendulum are two examples of objects that vibrate with shm. The conditions can be summarised by the equation: $a = -\omega^2 x$, where x represents displacement from an equilibrium position, a represents acceleration and ω is a constant that relates to the **period** of the vibration. The period, T, is the time taken to complete one oscillation; it is the inverse of the **frequency**, the number of oscillations per second, and is equal to $2\pi \div \omega$. An object vibrating with shm has displacement and velocity given by the equations: $x = A \sin \omega t = A \sin (2\pi f)t$ and $v = A\,\omega \cos \omega t = A\,(2\pi f)\cos 2\pi ft$ where A represents the amplitude of the vibration.

When **strings** vibrate each end is fixed and so forms a **node** (**no displacement**). The diagrams show the two longest wavelengths for a string vibration.

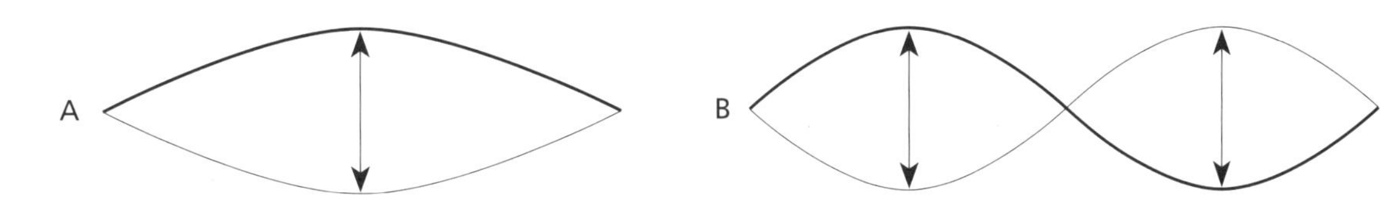

In diagram A the wavelength is twice the length of the string and the wavelength is equal to the length of the string in diagram B. The parts of the string that vibrate with the greatest amplitude (as shown by the arrows) are called **antinodes**. A vibrating **air column** has to have one end free, or open, where there is an antinode. The other end can be either closed or open; this allows a pipe to have more possible ways (or modes) of vibration than a string has.

Waves can be classified as either **transverse** or **longitudinal**. The oscillations in a transverse wave are at right angles to the direction of propagation of the wave whereas those in a longitudinal wave are parallel to it. Sound waves and compression waves are longitudinal; all electromagnetic waves are transverse. The wave speed equation velocity = frequency × wavelength, $v = f\lambda$ applies to all waves.

There is a change in the speed of a wave whenever the density of the material it is travelling in changes. This is known as refraction. The frequency of the wave motion is constant, and so a change in speed is accompanied by the same proportional change in wavelength. Unless the **wavefronts** are parallel to the boundary where the density change occurs, refraction results in a change in the direction of the wave.

The **absolute refractive index** of a material is defined by the equation $\mu = c_{vacuum} \div c_{material}$ where c represents the speed. The **relative refractive index** when a wave passes from one material to another is similarly defined as the ratio of the speeds, i.e. $\mu = v_1 \div v_2$. **Snell's law** relates the change in direction at refraction to the refractive index $\mu = \sin\theta_1 \div \sin\theta_2$, where θ_1 is the angle of incidence in material 1 and θ_2 is the angle of refraction in material 2.

REVISION SUMMARY

Total internal reflection can occur at a boundary where there is an increase in the wave speed. Waves that strike the interface at an angle greater than the **critical angle** are totally reflected. The critical angle can be calculated from the refractive index of a material using:

$$\sin c = \frac{1}{\mu}$$

A transverse wave is **polarised** when the oscillations are restricted to one plane that is perpendicular to the direction of travel. Radio waves that are transmitted from an aerial are polarised and light waves can become polarised when they pass through certain materials as well as becoming partially polarised when they are reflected.

Diffraction is the spreading out of waves when they pass through a gap or past an obstacle. The amount of observable spread when a wave passes through a gap depends on the size of the gap compared to the wavelength of the wave motion; the maximum spreading occurring when these are equal.

When two or more waves individually cause a displacement at a point then the total displacement is the algebraic sum (i.e. taking the directions into account) of the individual ones. This is known as the **principle of superposition** and it is used to explain **wave interference**.

Interference patterns from two sources are readily observable provided that the sources are **coherent**, which means they must have a fixed phase relationship. Two sources of sound or surface water waves which are **in phase** interfere **constructively** at any point where the **path difference** is equal to a whole number of wavelengths and **destructively** midway between these points, where the path difference is an odd number of half wavelengths. In order to show two-source interference with light it is necessary to use two narrow slits to produce 'copies' of a single light source by diffraction. These secondary sources may not be in phase but the phase difference between them does not change. The separation of the maxima produced by two-source interference is given by the formula $x = \lambda D/d$, where d is the separation of the sources and D is the distance from the sources to the point of observation.

A **diffraction grating** uses interference from many sources to create well-defined interference patterns. Constructive interference occurs when $n\lambda = d \sin \theta$, where d is the slit separation and n is the number of wavelengths in the path difference between waves from adjacent slits.

The **photo-electric** effect provides evidence that light travels in small units rather than as a continuous stream of waves. The energy of each unit, or photon, is hf, where h is Planck's constant. Each photon has a momentum $p = h/\lambda$. Similarly, some particles such as electrons can show wave-like properties such as diffraction and interference. The wavelength associated with a moving particle is called the de Broglie wavelength and is equal to h/p.

If you need to revise this subject more thoroughly, see the relevant topics in the *Letts A level Physics Study Guide.*

QUESTIONS

1 Which of the following is an example of resonance?

A A guitar string vibrates when it is plucked.
B The wing mirror of a lorry shakes violently when the lorry engine is idling.
C A tuning fork vibrates at a characteristic frequency when it is struck.
D A loudspeaker cone vibrates to reproduce a sound. (1)

2 A teacher can only hear sounds with frequencies between 30 Hz and 16 kHz. What is the shortest wavelength of sound she can hear if the speed of sound in air is 340 $m\,s^{-1}$?

A 1.13×10^{-2} m
B 2.13×10^{-2} m
C 8.82×10^{-2} m
D 11.3 m
E 47.1 m (1)

SEB

3 A beam of ultrasonic waves enters a liquid from air with an angle of incidence of 15° and continues with an angle of refraction of 60°. Given that the speed of ultrasonic waves in air is 330 $m\,s^{-1}$, what is their speed in the liquid?

A 82 $m\,s^{-1}$
B 99 $m\,s^{-1}$
C 1100 $m\,s^{-1}$
D 1320 $m\,s^{-1}$ (1)

Oxford

4 Monochromatic light, incident normally on a plane diffraction grating, may give rise to a number of transmitted beams. The diagram shows, as an example, a case where the total number of transmitted beams is three.

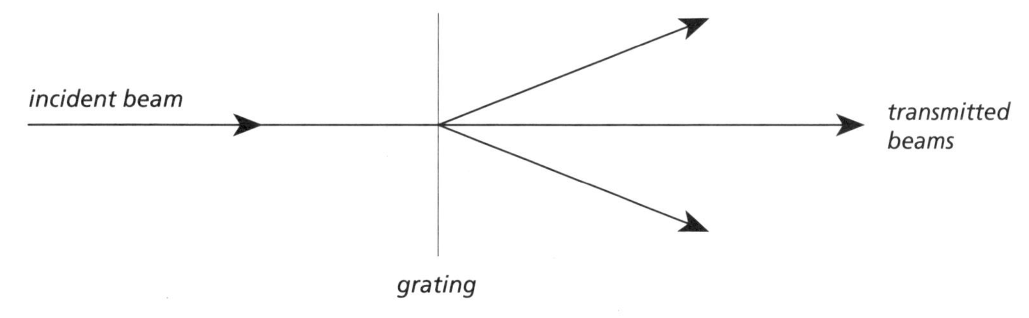

If the monochromatic light is of wavelength 6.3×10^{-7} m and the grating spacing is 20×10^{-7} m, what is the total number of transmitted beams?

A 3
B 4
C 5
D 6
E 7 (1)

NICCEA

5 Red light from a laser is passed through double slits. The diagram shows the pattern of dots produced on a screen.

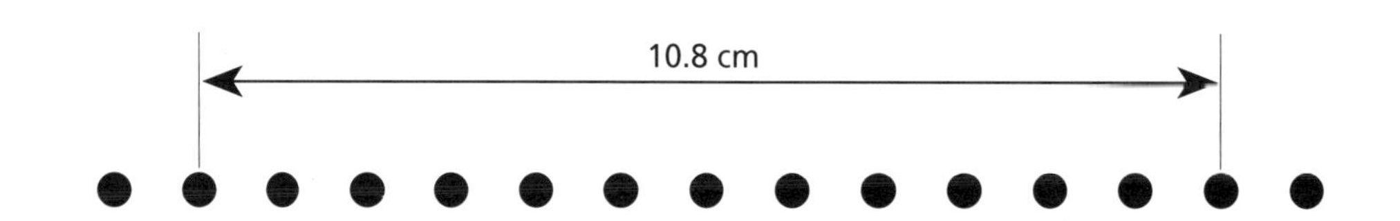

Other measurements are given below.

Distance from slits to screen = 3.02 m
Wavelength of laser light = 633 nm

(a) Using the expression $\lambda = \frac{xd}{D}$, calculate the separation of the slits.

....

....

(b) What change in the spacing of the dots will take place if a laser emitting green light is used instead?

....

.... (4)

SEB

6 An object moves along a straight line with simple harmonic motion of amplitude 10 cm and frequency $\frac{1}{\pi}$ Hz.

Sketch graphs to show the variation of:

(a) its acceleration with displacement,

(b) its displacement and velocity with time. (Plot both on the same time axis.) Insert appropriate numerical values along the axes of each sketch graph.

(8)

WJEC

7 (a) The diagram shows the path of a monochromatic beam of light through a triangular plastic prism.

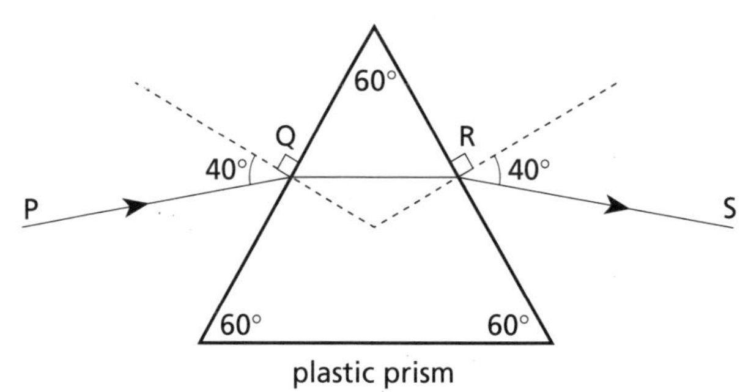

plastic prism

(i) Calculate the refractive index of the plastic.

....

....

(ii) Add to the drawing the path which the ray PQ would take from Q if the prism were made from a plastic with a **slightly higher** refractive index.

.... (3)

(b) The original prism is now replaced with one of the same size and shape but made from glass of refractive index 1.80.

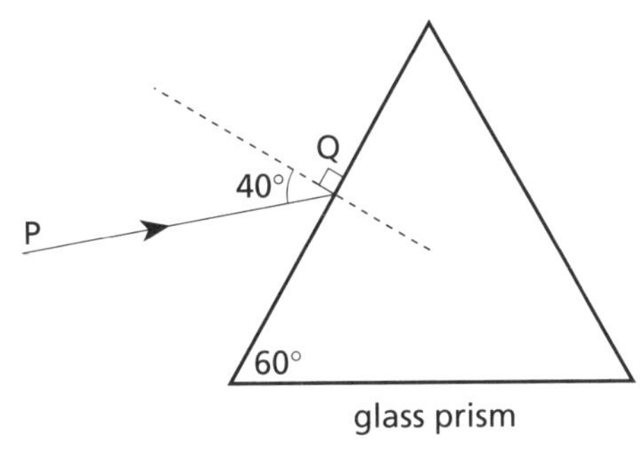

(i) Calculate the critical angle for this glass.

....

....

(ii) Draw on the diagram, showing the passage of the ray PQ through this prism until after it emerges into the air.
Mark on your diagram the values of all relevant angles. (5)

SEB

8 Two small loudspeakers, S_1 and S_2, are positioned 1.5 m apart in a large room and are connected to the same signal generator. The loudspeakers emit a note of 3400 Hz.

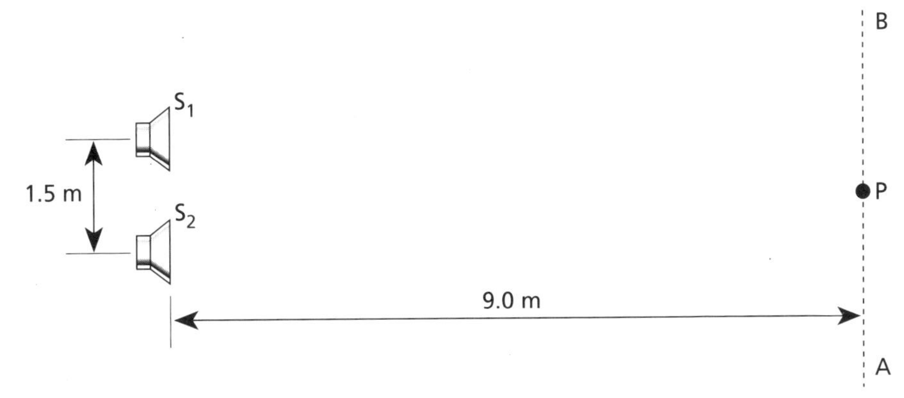

QUESTIONS

A microphone connected to an oscilloscope is placed at P which is equidistant from the two speakers as shown in the diagram. As the microphone is moved along the line AB, the trace height passes through a series of maxima and minima, with a maximum at P.
The speed of sound in air at room temperature = 340 m s^{-1}

(a) (i) Explain the variation in trace height (intensity).

....

....

....

....

(ii) Calculate the approximate distance from P to the first minimum.

....

....

....

.... (5)

(b) Explain how you would expect the distance you have calculated in (a) to change if the temperature of the air in the room falls, assuming that the speed of sound waves decreases with decrease in temperature.

....

....

....

.... (2)

NEAB

9 (a) In the photoelectric effect, electrons may be emitted from a metal when electromagnetic radiation falls on the surface. Monochromatic radiation from source A is shone in turn on metal P and then on metal Q. This procedure is repeated with source B and then source C. You may assume that the apparatus is arranged so that emission of electrons can be detected. The results are summarised in the table below.

	wavelength/nm	metal P	metal Q
Source A	300	electrons emitted	electrons emitted
Source B	600	electrons emitted	no electrons emitted
Source C	800	no electrons emitted	no electrons emitted

(i) Visible light has a range of wavelengths from 400 to 700 nm. State in which region of the electromagnetic spectrum lies each of the radiations from source A and from source C.

....

....

(ii) By considering the Einstein photoelectric equation, explain the pattern of the results obtained.

....

....

....

....

....

(iii) How would you expect the results to be different, if at all, if the light from source B is doubled in intensity (i.e. is made twice as bright)? Explain your answer.

....

....

....

.... (9)

(b) Electrons are known to show wave properties, as illustrated by electron diffraction.

(i) When electrons are accelerated from rest by a potential difference, V, to a speed, v, in a vacuum, these quantities are related by the equation

$$\frac{1}{2}mv^2 = eV,$$

where e is the charge on the electron and m is its mass.
Explain how the equation is an application of the principle of conservation of energy.

....

....

....

(ii) Calculate the speed of electrons accelerated from rest through a p.d. of 2000 V.
electron charge, $e = 1.60 \times 10^{-19}$ C
electron mass, $m = 9.11 \times 10^{-31}$ kg

....

....

....

.... (4)

(c) In a laboratory demonstration of electron diffraction, electrons are accelerated in a vacuum tube and pass through a thin disc of graphite on to a fluorescent screen at the end of the tube, where a pattern of concentric rings is seen.

(i) Use the momentum-wavelength equation to show that the wavelength associated with the electrons referred to in (b)(ii) above is 2.7×10^{-11} m.
Planck's constant, $h = 6.6 \times 10^{-34}$ J s

....

....

....

(ii) By reference to your answer in (c)(i), explain why electrons can be diffracted through quite large angles by passing them through a thin sheet of graphite.

....

....

....

....

(iii) Describe one simple test you could carry out using the above apparatus to support the idea that these rings are produced by beams of **negatively charged** particles. Explain, with the aid of a diagram, how you would reach this conclusion from the observations made.

....

....

....

....

(7)
NEAB

10 (a) The refractive index n for light passing from air to glass is given by the expression $n = \sin i / \sin r$.

(i) Draw a diagram to illustrate this relationship. Identify and mark clearly the angles i and r.

(3)

(ii) Explain what is meant by *total internal reflection*.

....

....

.... (2)

QUESTIONS

(iii) Show, with the aid of a diagram, that the critical angle θ for total internal reflection at a glass-air interface is given by

$$\sin\theta = \frac{1}{n}$$

....

....

.... (3)

(b) The diagram shows a length of homogenous glass fibre of refractive index n and radius R bent into a circular arc of mean radius S. Such fibres are used for 'light-piping' and for the transmission of optical data pulses.
A parallel beam of monochromatic light enters the fibre normally at the left-hand plane face.

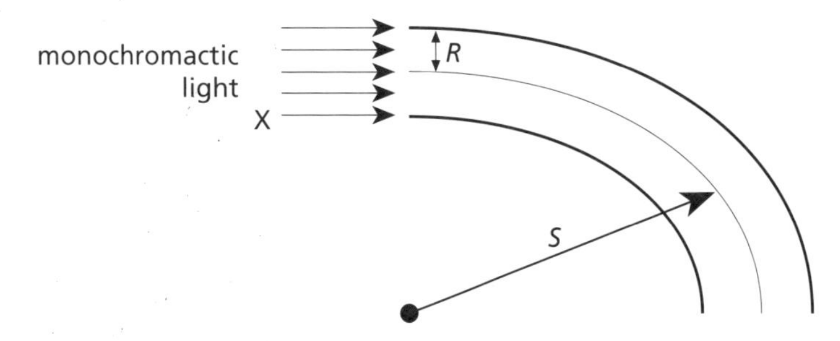

(i) Show the path within the fibre of X, the lowest ray entering the left-hand face of the fibre. (2)

(ii) Use your diagram from (i) and the critical angle relation from (a)(iii) to show that some light could escape from the fibre if its refractive index n had a value given by

$$n < \frac{S+R}{S-R}$$

....

....

....

....

....

.... (5)

QUESTIONS

(c) Data pulses can be transmitted from place to place by microwave (radio) links, by light pulses through glass fibre, or by electric currents in copper cable.
Take the speed of electromagnetic waves in free space as 3.0×10^8 m s^{-1} and the refractive index of glass as 1.5.

(i) Calculate the minimum difference in transmit times over a 50 km distance using microwave and light-pulse systems.

....

....

....

.... (3)

(ii) Describe a situation in which glass fibre would be preferred to microwaves. Give a technical reason for this preference.

....

....

.... (2)

Oxford

4 Gravitational and electrical fields

REVISION SUMMARY

A **field** describes a region of space where forces are exerted on objects with certain properties. **Gravitational fields** affect objects that have mass and **electric fields** affect those with charge. Magnetic fields are considered in Unit 6.

Gravitational and electric fields have important similarities and differences; the key difference is that whilst gravitational forces are always attractive, electric forces can be attractive or repulsive. The convention is that electric field lines are drawn with the arrows showing the direction of the force on a positive charge. The diagrams show (a) the electric field due to a point (positive) charge and (b) the gravitational field due to a point mass.

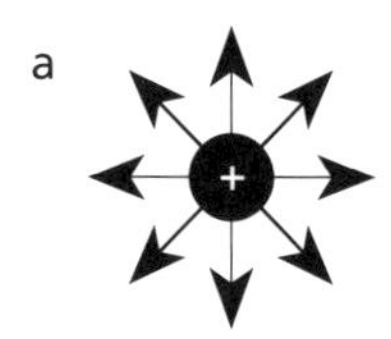

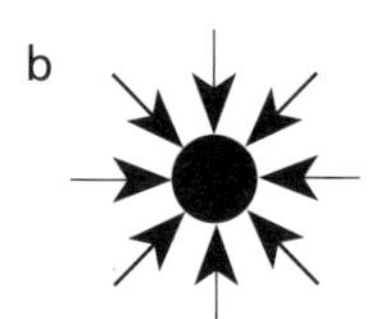

These field lines radiate in three dimensions, so at double the distance from the centre of the charge or mass the field lines are spread over four times the area. This explains why **Coulomb's law** for the force between two charges and **Newton's law** for the force between two masses are **inverse square** laws. They are summarised by the equations

$$F = \frac{1}{4\pi\varepsilon} \times \frac{q_1 q_2}{r^2} \quad \text{and} \quad F = G\frac{m_1 m_2}{r^2}.$$

The constant ε stands for the permittivity of the medium, where ε_0 is used for free space or a vacuum, and G is the universal gravitational constant.

The **electric field strength**, E, at a point in an electric field is defined as being the force per unit charge at that point. Similarly, **gravitational field strength**, g, is defined as the force per unit mass. It follows from Coulomb's and Newton's laws that in radial fields due to a charge Q or a mass M:

$$E = \frac{F}{q} = \frac{1}{4\pi\varepsilon} \times \frac{Q}{r^2} \quad \text{and} \quad g = \frac{F}{m} = G\frac{M}{r^2}.$$

Not all fields are radial; the electric field between a pair of charged flat plates is uniform and is given by the formula

$$E = \frac{V}{d}$$

where V is the potential difference between the plates and d is their separation. The gravitational field close to the Earth's surface is considered to be uniform as there is little change in g up to a height of several thousand metres.

The **potential** at a point in a gravitational or an electric field measures the potential energy that each unit of mass or charge would have if placed at that point. The absolute potential is measured relative to infinity, where the potential energy is defined as zero. Potentials in gravitational fields are always negative because a mass loses potential energy in moving from infinity to a point in the field as the attractive gravitational force does work on the mass. The electric potential in a field due to a positive charge is positive because another positive charge would gain energy in moving from infinity. The formulas for gravitational and electric potential in a field due to a point mass or positive charge are:

$$V_g = -\frac{GM}{r} \quad \text{and} \quad V_E = \frac{Q}{4\pi\varepsilon r}.$$

The **potential difference** between two points in a field is the energy transfer for each unit of mass or charge that is moved between the points, $V_g = W/M$ or $V_E = W/Q$.

In both gravitational and electric fields field strength and potential are linked by the relationship *field strength = – (potential gradient).*

If you need to revise this subject more thoroughly, see the relevant topics in the *Letts A level Physics Study Guide.*

Letts Q&A

QUESTIONS

1 If F is the gravitational force between two particles and d is their separation, which of the following graphs best represents the relationship between F and d?

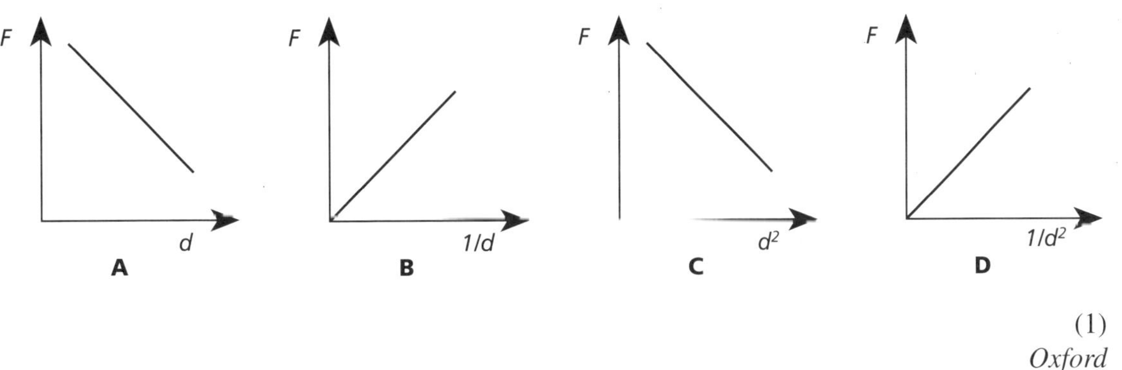

(1)
Oxford

2 A satellite is in orbit where the gravitational potential is -50 MJ kg^{-1}. It moves to a higher orbit. The potential in the higher orbit could be:

A -60 MJ kg^{-1}
B -40 MJ kg^{-1}
C 0 MJ kg^{-1}
D $+20$ MJ kg^{-1} (1)

3 A negatively charged particle is held stationary between two horizontal charged conducting plates, as shown in the diagram.

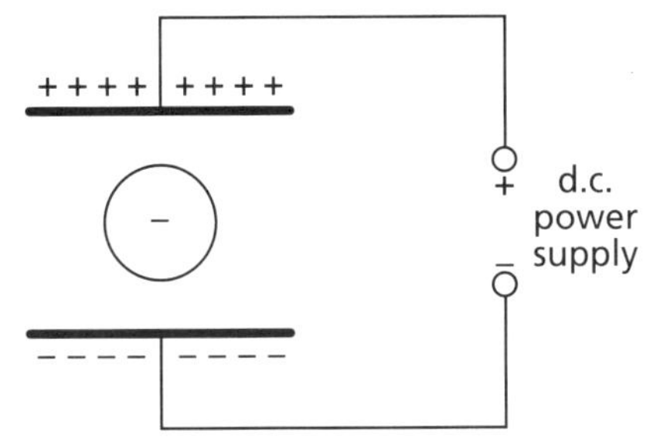

Which of the following changes would cause the particle to move downwards?

A Increase the electric field strength.
B Increase the charge on the particle.
C Change the position of the particle.
D Decrease the mass of the particle.
E Decrease the electric field strength. (1)

SEB

4 The electric potential V at a point on the straight line joining two points P and Q varies with distance x as shown in the diagram.

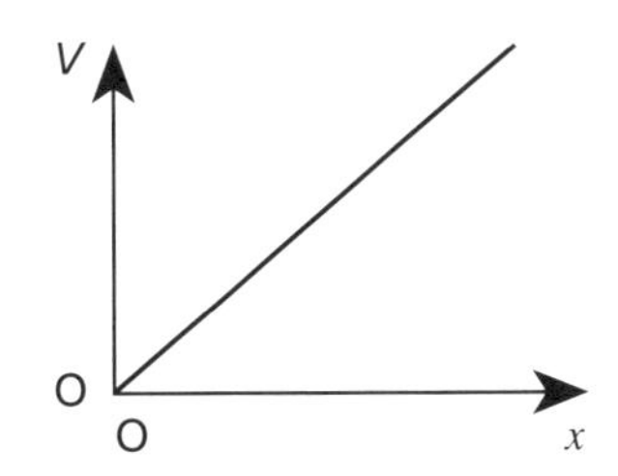

Which one of the graphs below correctly shows the variation with x of the electric field strength, E, directed along PQ? (1)

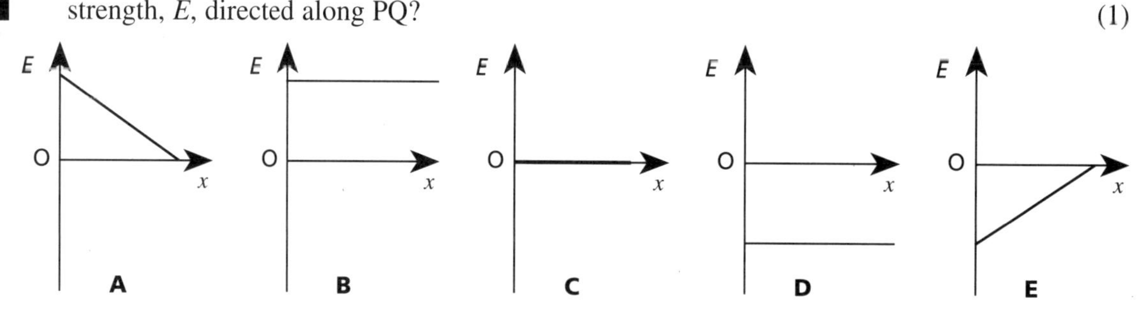

NICCEA

5 The rings of the planet Saturn consist of a vast number of small particles, each in a circular orbit about the planet. Two of the rings are shown in the diagram.

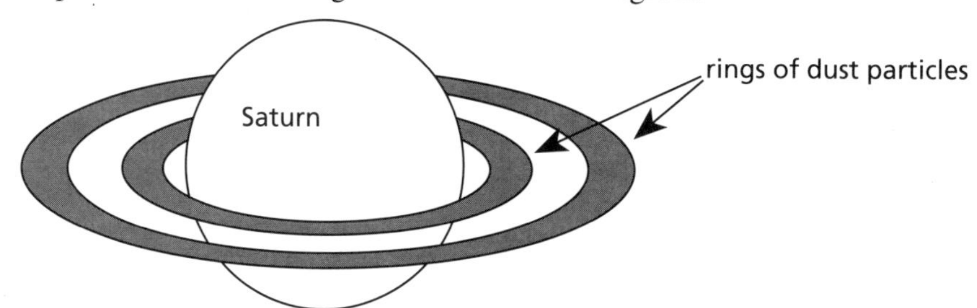

The inner edge of the inner ring is 70 000 km from the centre of the planet and the outermost edge of the outer ring is 140 000 km from the centre. The speed of the outermost particles is 17 km s^{-1}.

(a) Show that the speed, v, of a particle in an orbit of radius r around a planet of mass M is given by

$$v = \sqrt{\frac{GM}{r}}$$

where G is the universal gravitational constant, 6.7×10^{-11} N m^{2}kg^{-2}.

....

....

....

(b) Determine the mass of Saturn.

....

....

....

QUESTIONS

(c) How long does it take for the outermost particles to complete an orbit?

....

....

....

(d) Calculate the orbital speed of the particles nearest to Saturn.

....

....

.... (7)

AEB

6 The Earth may be regarded as a uniform sphere of mass M and radius r.

(a) (i) Write down expressions for the force experienced by a mass m on the Earth's surface in terms of:
m and the gravitational field strength g;

.... (1)

m, M, r and the gravitational constant G.

....

.... (1)

(ii) Hence obtain an expression for M in terms of G, g, and r. (1)

....

(b) The Earth's mean radius is 6400 km, and the value of G is 6.7×10^{-11} N m^{2}kg^{-2}.
[Volume of sphere of radius $r = (4/3)\pi r^3$.]
Calculate a value for the mean density of the Earth.
[Take $g = 9.81$ m s^{-2}]

....

....

.... (3)

QUESTIONS

(c) Measurements at a scientific station at the North pole show that the value of g, the acceleration of free-fall, is slightly greater than the value elsewhere in the northern hemisphere. Suggest a reason why. Assume there are no significant variations in the density of the Earth.

....

....

.... (2)

Oxford

7 Two parallel plates have a potential difference of 200 V across them. Draw a diagram to show the shape of the electric field between the plates. Add to the diagram equipotentials at 50 V and 100 V. (Assume the plates are identical and co-terminous.)

(5)

ULEAC

8 (a) (i) Define *electric potential.*

....

....

....

(ii) The electric potential V at a distance r from a point charge Q in vacuo (free space) is given by

$$V = \frac{Q}{4\pi\varepsilon_0 r}$$

Write down an expression for the electric potential energy of a point charge q when at a distance r from a point charge of Q in vacuo (free space).

....

(b) An α-particle of charge $+3.2 \times 10^{-19}$ C and mass 6.8×10^{-27} kg is travelling at a speed of 1.2×10^{7} m s^{-1} directly towards a fixed nitrogen nucleus of charge $+11.2 \times 10^{-19}$ C. Assuming that initially they are far apart calculate the closest distance of approach.

[Take $\frac{1}{4\pi\varepsilon_0} = 9.0 \times 10^{9}$ F^{-1}m]

....

....

....

....

....

.... (8)

WJEC

9 (a) The gravitational field strength, g, at the Earth's surface is 9.8 N kg^{-1}. The radius of the Earth is R.
Draw a labelled sketch graph to show how the gravitational field strength, g, varies with h, the height above the Earth's surface, over a range $h = 0$ to $h = 2R$.

(b) A 40 000 kg spacecraft moves away from the Earth with its motors switched off.
The table shows two corresponding values for the kinetic energy, E_K, of the spacecraft and its distance, r, **from the centre of the Earth.**

Position	$E_K/10^{10}$ J	$r/10^6$ m
A	57.80	26.30
B	52.10	29.03

(i) Calculate the speed of the spacecraft at position **A**.

....

....

(ii) Calculate the work done by the spacecraft between **A** and **B**.

....

.... (7)

AEB

10 (a) Define the term *gravitational field strength.*

....

.... (1)

QUESTIONS

(b) State the numerical value and the unit of the gravitational field strength of the Earth at its surface.

.... (2)

(c) Why is it incorrect to call g (= 9.8 m s^{-2}) 'gravity'?

....

....

.... (2)

(d) This part of the question is about the rotation of the Moon in a circular orbit around the Earth. You will need to use the following astronomical data.

Radius of the Moon's orbit	$= 3.84 \times 10^8$ m
Mass of the Moon	$= 7.35 \times 10^{22}$ kg
Time for Moon to complete one orbit around the Earth	$= 2.36 \times 10^6$ s

Calculate

(i) the speed of the Moon in its orbit around the Earth,

....

....

(ii) the acceleration of the Moon,

....

....

(iii) the force the Earth exerts on the Moon,

....

....

(iv) the gravitational field strength of the Earth at the Moon.

....

.... (6)

UCLES

The job of a circuit is to transfer energy from the source to the components. This transfer is done by the charge carriers that move around the circuit. In the case of metallic conductors these charge carriers are free electrons. The **current** in a circuit is defined as being the rate of flow of charge, $I = Q/t$. One factor that affects the current in a circuit is the **resistance**, which is defined by the formula $R = V/I$. The resistance of an object depends on its physical dimensions as well as the **resistivity**, ρ, of the material itself; if these are known the resistance can be calculated from the formula

$$R = \frac{\rho l}{A}.$$

Ohm's Law states that the current is directly proportional to the voltage applied i.e. the resistance of a metallic conductor stays constant, provided that the temperature and other physical conditions also remain constant. Increasing the temperature causes the resistance of metallic conductors to become greater but has the opposite affect on semiconductors. A **diode** is a device that allows current to pass in one direction only.

Potential difference measures the energy transfer, W, for each coulomb of charge passing between two points, i.e.

$$V = \frac{W}{Q}.$$

The potential difference across a device such as a resistor measures the energy transfer from each coulomb of charge to the resistor. The total energy transfer to each coulomb of charge by a power supply or battery is called the **electromotive force** or **e.m.f**. When the power supply is driving a current the potential difference across the terminals is less than this because of its **internal resistance**. This acts like a series resistor in the circuit and results in some of the available energy being transferred within the power supply itself. The potential difference across the power supply is given by the formula $V = E - Ir$, where E is the emf and r is the internal resistance of the power supply.

Power, or the rate of energy transfer, is calculated using the formula $P = IV$ or the equivalent form, $P = I^2R$.

When two or more resistors are placed in **series** the effect is to increase the total resistance, $R_S = R_1 + R_2 + R_3$. Two or more resistors in **parallel** have a combined resistance that is smaller than that of the lowest value resistor, $\frac{1}{R_P} = \frac{1}{R_1} + \frac{1}{R_2} + \frac{1}{R_3}$. Series resistors are used in a **potential divider**; potential dividers are used in electronics to derive a smaller p.d. than that of the available supply. Where one of the resistors used is sensitive to environmental conditions e.g. temperature, light intensity etc., potential dividers provide a varying p.d., which can be used to switch mains devices such as lamps, heaters and motors.

Kirchoff's laws apply laws of conservation to circuits. The first law states that *the algebraic sum of the currents at a junction is zero*. This means that the current passing into a junction is equal to the current passing out of it, a restatement of the law of **conservation of charge**. The second law states that *around any closed circuit loop, the gain in e.m.f. is equal to the loss in p.d.* This means that a certain **potential** is associated with any point in a circuit, so that in moving from that point around a loop and back again, the energy transferred to a charge (from a source) is equal to the energy transferred by it; an application of the principle of **conservation of energy** to a circuit.

REVISION SUMMARY

Electric charge is stored in a **capacitor**. The capacitance of a capacitor, defined by the equation

$$C = \frac{Q}{V},$$

measures the charge stored for each volt of potential difference. For a parallel plate capacitor, the capacitance can be calculated using the formula

$$C = \frac{\varepsilon A}{d},$$

where A represents the area of overlap and d represents the separation of the capacitor plates. As a capacitor is discharged through a resistor, the potential difference falls and so does the discharge current. This leads to an **exponential decrease** of the charge on the capacitor. The rate at which the charge flows from the capacitor is governed by the time constant, RC. This represents the time it takes for the charge on the capacitor to decrease to $\frac{1}{e}$ of its initial value, where e is the base of natural logarithms. The time taken for the charge to fall to one half of its initial value is $0.69RC$.

A similar situation exists when a capacitor is charging through a resistor, except this time the resultant p.d. that causes the current is the difference between the supply p.d. and that across the capacitor. As the capacitor charges, this resultant p.d. falls, so the charging current also falls. The time constant RC represents the time it takes for the charge to rise to within $\frac{1}{e}$ of its final value, the time taken to rise to one half of its final value being $0.69RC$.

The gradient of a charge-time graph for a capacitor represents the current and the area between the line and the time axis of a current-time graph represents the charge stored.

The effect of placing capacitors in **series** is to decrease the capacitance, $\frac{1}{C_S} = \frac{1}{C_1} + \frac{1}{C_2} + \frac{1}{C_3}$, whilst placing capacitors in parallel increases the charge stored and so increases the capacitance, $C_P = C_1 + C_2 + C_3$.

The energy stored in a charged capacitor can be calculated from either of the three equivalent expressions $E = {}^1/_2QV = {}^1/_2CV^2 = {}^1/_2Q^2/C$.

If you need to revise this subject more thoroughly, see the relevant topics in the Letts *A level Physics Study Guide.*

QUESTIONS

1 The reading on the high resistance voltmeter in the circuit shown below is 1.0 V.

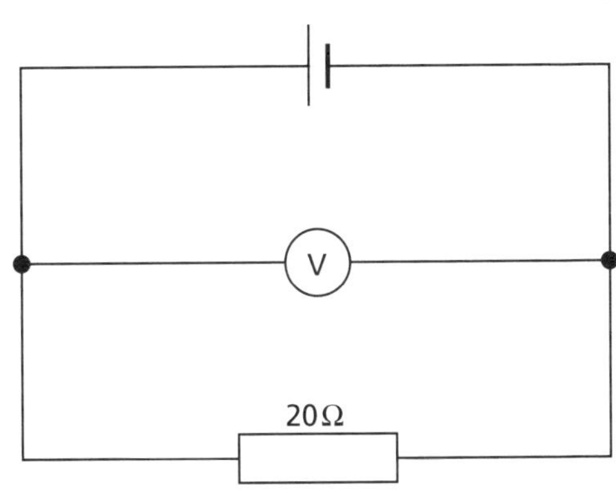

The e.m.f. of the cell is 1.5 V.
The internal resistance of the cell is

A 0.1 Ω
B 0.5 Ω
C 1.0 Ω
D 2.5 Ω
E 10 Ω

(1)
SEB

2 A battery is connected to three resistors in series.

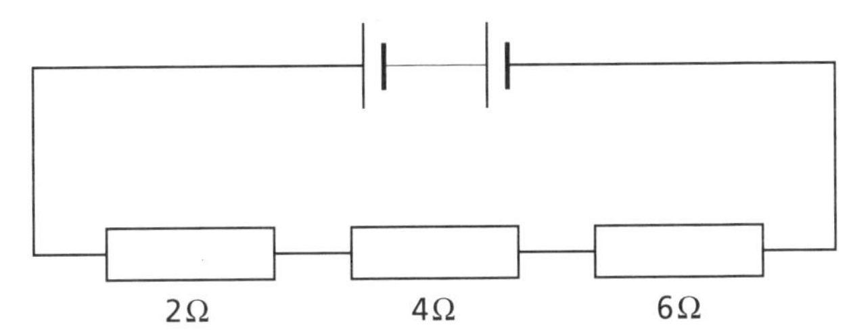

The power supplied by the battery is 0.12 W. What is the current in the circuit?

A 0.01 A
B 0.06 A
C 0.10 A
D 0.20 A

(1)

3 Four capacitors are connected as shown in the diagram.

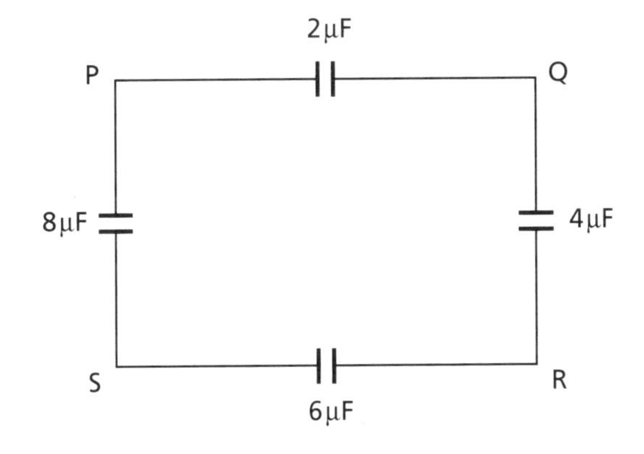

QUESTIONS

Between which of the following pairs of points is the maximum capacitance of the combination obtained?

A P and Q
B Q and R
C R and S
D P and S
E Q and S

(1)
NICCEA

4 A 10 μF capacitor is fully charged using a 12 V supply. Which **one** of the statements correctly gives the charge on each plate of the capacitor and the energy it stores?

	charge/C	**energy/J**
A	1.2×10^{-4}	6.0×10^{-5}
B	1.2×10^{-4}	7.2×10^{-4}
C	1.2×10^{6}	7.2×10^{-4}
D	1.2×10^{6}	6.0×10^{-10}

(1)
AEB

5 A car driver parks his car with two 60 W headlamps and four 6 W sidelamps still switched on. The six lamps are connected in parallel and powered by a 12 V battery.
Explain the phrase *connected in parallel.*

....

....

....

Calculate the current in the battery.

....

....

....(6)
ULEAC

6 A heater of resistance 0.32 ohms is connected to a power supply of e.m.f. 2.0 volts and internal resistance r as shown below.

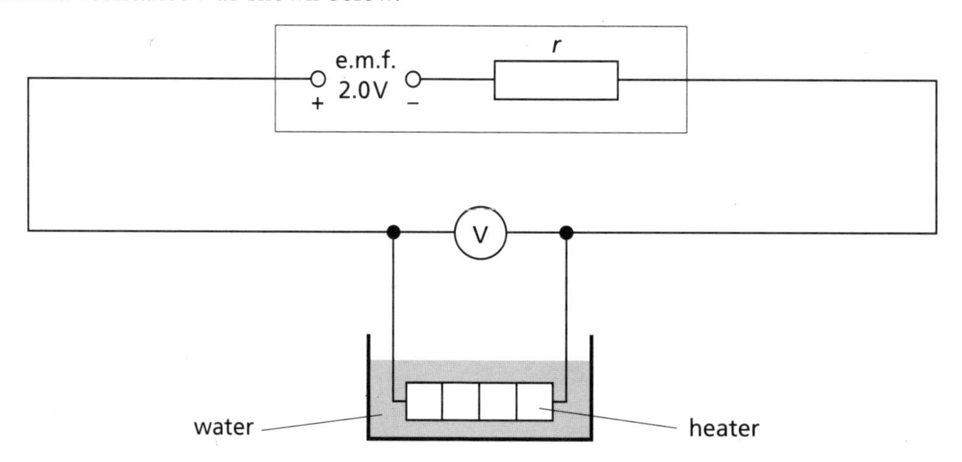

(a) State what is meant by the term electromotive force (e.m.f.).

....

.... (1)

(b) The power output of the **heater** is 8.0 watts.
Calculate:

(i) the current in the heater;

....

....

(ii) the reading on the voltmeter;

....

....

(iii) the internal resistance of the power supply.

....

.... (5)

(c) Another identical heater is now placed in the water and connected in parallel with the original heater.
The rest of the circuit is unaltered.
How does this affect the rate at which heat is supplied to the water?
Justify your answer by calculation.

....

....

.... (3)

SEB

7 (a) A workshop produces parallel plate capacitors of the type shown below. Air is the insulator between the plates.

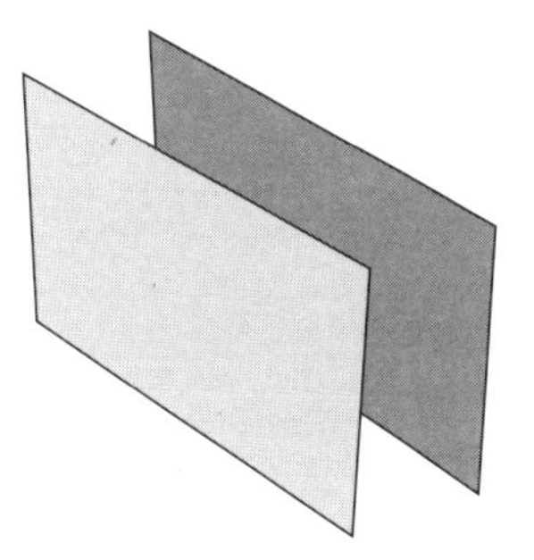

QUESTIONS

The area of the plates can be made either A or $2A$.
The separation of the plates can be made either d or $2d$.

(i) Describe how the capacitor of smallest capacitance would be made.

....

....

(ii) A slab of polythene is now put between the plates of this capacitor.
What is the effect on the capacitance?

....

(iii) One of the capacitors from the workshop was tested and the following graph was obtained.

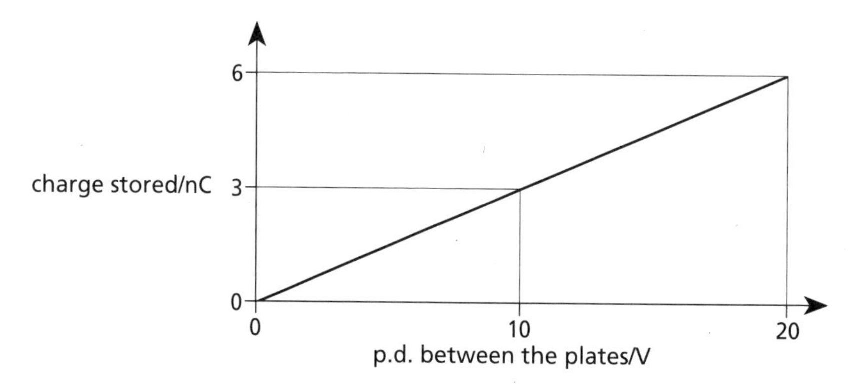

Calculate:

(A) the capacitance of the capacitor;

....

....

(B) the energy stored by the capacitor when charged to a p.d. of 10 V;

....

....

(C) the energy stored by the capacitor when charged to a p.d. of 20 V.

....

.... (7)

(b) A parallel plate capacitor has the following information written on it:

'400 V maximum working'

(i) Explain why the capacitor has a maximum working voltage.

....

(ii) The medium between the plates of this parallel plate capacitor ceases to be an effective insulator when it is subject to an electric field strength of 30 kV cm^{-1}. What is the minimum spacing of the plates?

....

.... (3)

(c) In the circuit shown below, the capacitor is initially uncharged, and the battery has negligible internal resistance.

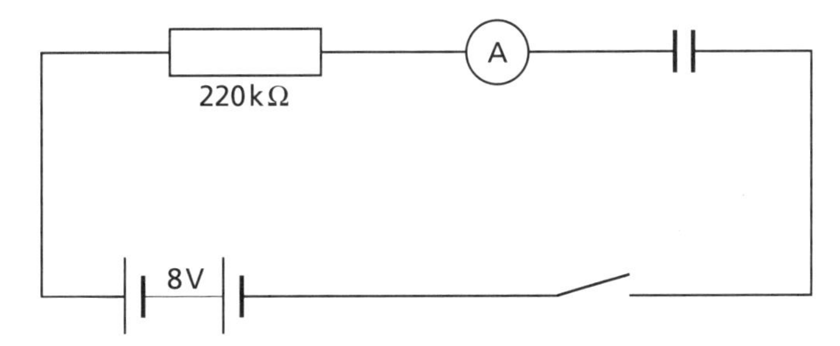

(i) Calculate the current immediately the switch is closed.

....

....

(ii) (A) Sketch a graph to show how the p.d. across the capacitor changes with time after the switch is closed.

(B) The 220 kΩ resistor is replaced by a 470 kΩ resistor and the experiment is repeated.
On the same axes, sketch the new graph of p.d. against time. Put the label '470 kΩ' on the new graph.

QUESTIONS

(iii) The 8 V battery is replaced with an 8 V a.c. supply. The ammeter is changed to its 'a.c.' setting and the switch is closed.
What change occurs in the ammeter reading if the frequency of the alternating supply is increased?

....

.... (5)

SEB

8 (a) (i) A capacitor consists of two parallel metal plates separated by an insulator. A d.c. power supply of variable output voltage may be connected across the plates. When a voltage is applied between the plates, the plates become charged. Explain briefly, in terms of the movement of charges, why the plates become oppositely charged.

....

....

Sketch a graph showing how the magnitude of the charge on a plate alters as the supply voltage is increased. How may the value of the capacitance of the capacitor be obtained from such a graph obtained by experiment?

....

....

Write down an expression for the energy stored in a capacitor of capacitance C charged to a potential difference of V. Suggest an explanation for the fact that there is a limit to the quantity of energy that may be stored in a practical capacitor, even though a power supply of unlimited voltage may be available.

....

....

....

.... (10)

(ii) Draw a labelled diagram of the circuit used to determine the capacitance of a capacitor by the vibrating reed experiment. What results would you record in such an experiment, and how would you use them to deduce the value of the capacitance?

....

....

....

....

.... (7)

(b) The diagram shows an arrangement which may be used to charge a capacitor of capacitance 20 μF, and then to connect it to a capacitor of capacitance 10 μF.

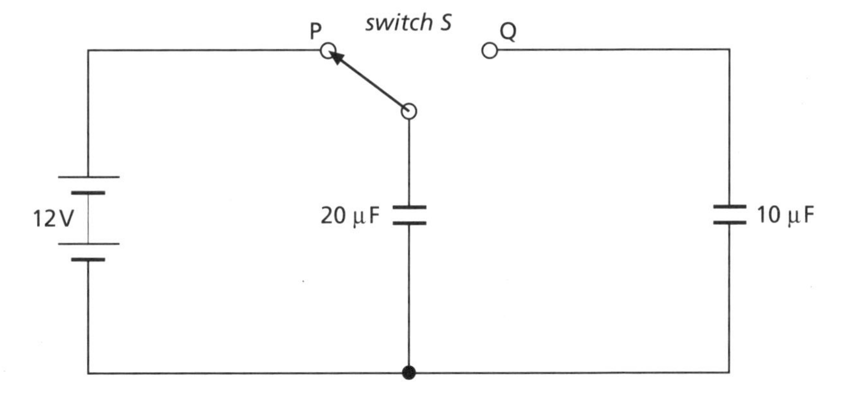

(i) The switch S is first placed at position P, so that the capacitor of capacitance 20 μF is connected to the 12 V d.c. supply. Calculate the energy stored in the capacitor when it is charged in this way.

....

.... (2)

QUESTIONS

(ii) The switch S is now changed to position Q. Calculate the final potential difference across the capacitors.

....

....

....

.... (4)

(iii) Calculate the difference between the final total energy stored in both capacitors and the initial energy stored in the capacitor of capacitance 20 μF. Suggest an explanation for this difference.

....

....

....

....

....

.... (5)

NICCEA

REVISION SUMMARY

Magnetic fields can be due to permanent magnets or electric currents and they exert forces on both of these. The **direction** of a magnetic field is always shown as the direction in which a force would be exerted on the north-seeking pole of a magnet. The **strength** of a magnetic field, B, is defined as the force exerted per unit (current × length) that is perpendicular to the field; i.e. $B = F/Il$. The mutually perpendicular directions of the magnetic field, the current and the force are given by the **left hand rule**.

This definition leads to the formula $F = BIl$ for the force on a current-carrying conductor and $F = Bqv$ for the force on a moving charge, provided that these are perpendicular to the field direction. Otherwise they apply to the component of the vectors Il and qv that is perpendicular to the magnetic field.

The magnetic field due to a **current** in a straight conductor can be represented as concentric circles around the current. The direction is clockwise when viewed in the same direction as the current and the field strength at a distance a from the current is given by $B = \mu I/2\pi a$, where μ represents the **permeability** of the medium. Iron has a high permeability, but air is normally considered to have the same permeability as a vacuum, μ_0. Inside but away from the ends of a **solenoid**, the uniform field has a constant value of $B = \mu nI$, where n represents the number of turns of wire per metre.

The force between two parallel, current-carrying conductors, $F = \mu I_1 I_2 l/2\pi a$, is used as the basis of the definition of the **ampère**.

The term **magnetic flux density** is an equivalent term to magnetic field strength. It refers to a model of magnetic fields being due to **flux**. On this model the amount of flux, φ, passing through an area at right angles to the flux is given by the formula $\varphi = BA$. A coil of n turns at right angles to a magnetic field is said to have a **flux-linkage** $\Phi = n\varphi = nBA$.

The flux model is useful when considering **electromagnetic induction**. An e.m.f. is **induced** in a conductor whenever it is subjected to a change in magnetic field. This change can arise due to movement of the conductor, e.g. an aircraft flying through the Earth's magnetic field, or movement of the field, e.g. a magnet rotating inside a coil of wire in a dynamo, or a changing current, e.g. the alternating current in a coil of a transformer.

The induced e.m.f. causes a current to pass if there is a complete circuit.

Faraday's law states that the size of the induced e.m.f. is equal to the rate of change of flux-linkage. For a uniform change in flux-linkage this can be written as

$$\xi = \frac{\Delta\Phi}{\Delta t}.$$

Lenz's law states that the direction of the induced e.m.f. is always such as to oppose the change that causes it. This can be illustrated by the induced e.m.f. produced when a magnet is pushed into a coil of wire. It causes a current with a magnetic field that repels the magnet. Faraday's law and Lenz's law can be combined together into the single equation

$$\xi = -\frac{d\Phi}{dt},$$

or

$$\xi = -\frac{\Delta\Phi}{\Delta t}$$

if the flux-linkage is changing at a uniform rate.

The induced emf in an alternator or generator varies with time as the rate of change of flux linkage is **non-uniform**. The instantaneous value is given by the formula $E = E_0 \sin 2\pi ft$, where E_0 is the **peak** value of the induced e.m.f. given by the formula $E_0 = BAN\omega$.

REVISION SUMMARY

Transformers use electromagnetic induction to transfer energy from a **primary** coil to a **secondary** coil. In transferring this energy, the transformer can change the size of an alternating current and voltage. Transformers are very efficient and it is usual to neglect the small amount of energy transferred to thermal energy. This leads to the 'ideal' transformer equations that the voltage ratio is equal to the turns ratio,

$$\frac{V_p}{V_s} = \frac{n_p}{n_s}$$

and the current ratio is the inverse of this,

$$\frac{I_p}{I_s} = \frac{n_s}{n_p}$$

Transformers are used extensively in the national grid network to minimise energy losses as thermal energy in the transmission cables. Because this energy transfer is proportional to the $(\text{current})^2$, it is necessary to minimise the current by transmitting power at high voltages.

The use of transformers in the distribution of electrical energy means that **alternating current** has to be used. The effective, or root mean square, value of an alternating current is defined as being the value of the direct current that has the same heating effect. For a sinusoidally varying a.c. such as the mains supply, this is given by the formula $I_{rms} = \frac{I_{peak}}{\sqrt{2}}$. Similarly, $V_{rms} = \frac{V_{peak}}{\sqrt{2}}$. When these are combined together to give the power transfer, the rms power is equal to half the peak power, $P_{rms} = \frac{P_{peak}}{2}$.

If you need to revise this subject more thoroughly, see the relevant topics in the *Letts A level Physics Study Guide.*

1 **X**, **Y** and **Z** are three coplanar long parallel wires carrying currents as shown.

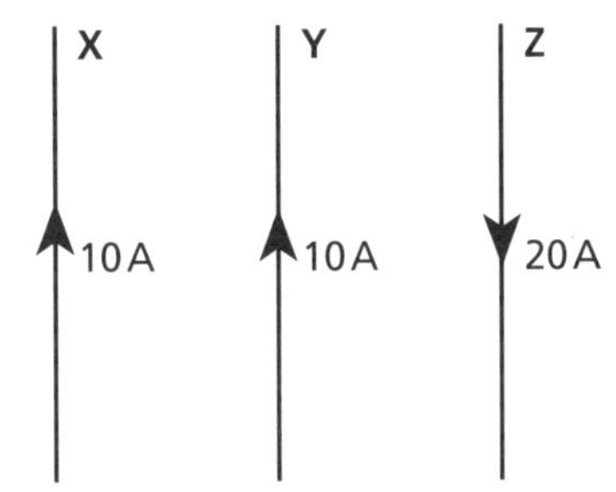

The resultant force on wire **Y** is

A zero
B towards **X**
C towards **Z**
D perpendicular to the paper

(1)
AEB

2 A plotting compass is placed next to a vertical wire. When there is no current in the wire the compass points due north.

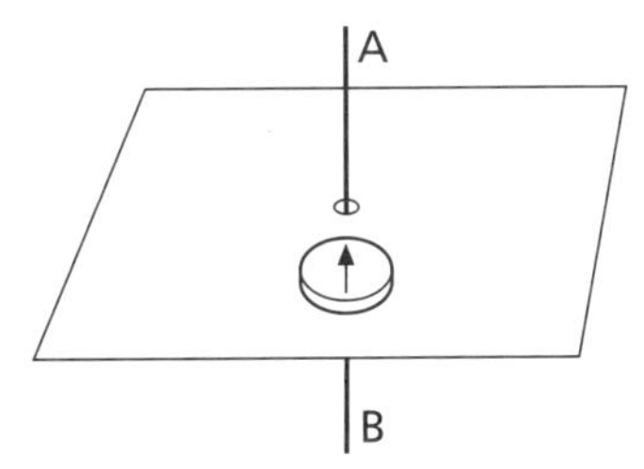

Which diagram shows a possible direction for the compass to point when a current passes from A to B?

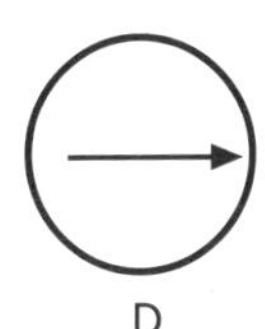

(1)

3 The diagram shows an oscilloscope trace obtained by connecting an alternating potential difference to the Y input of an oscilloscope.

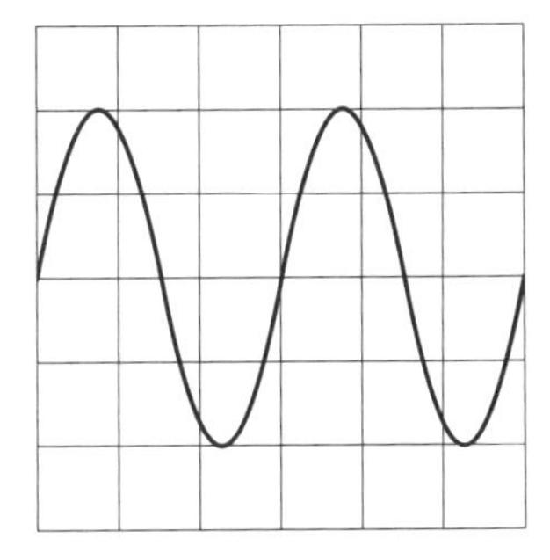

The grid is made up of 1 cm squares.
The Y-sensitivity is set to 10 V cm^{-1}.

QUESTIONS

The r.m.s. potential difference is

A 40 V
B 28 V
C 20 V
D 14 V

(1)

4 The following transformer has a turns ratio of 1:4.

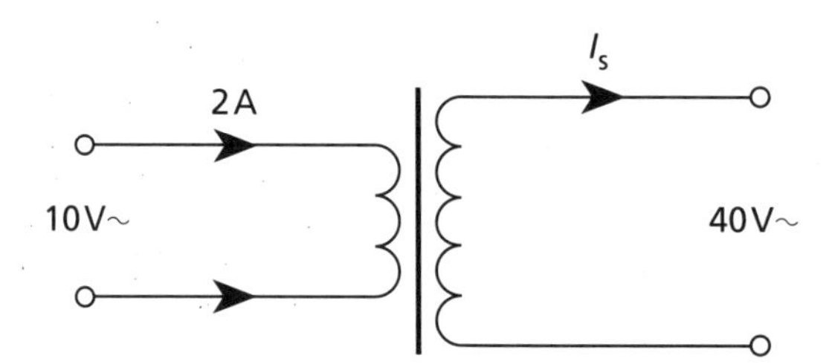

If electrical energy is converted to heat in the windings and the core of the transformer at a rate of 4 W, what is the value of the secondary current I_s ?

A 0.1 A
B 0.4 A
C 0.5 A
D 0.6 A
E 0.8 A

(1)
SEB

5 (a) The diagrams below show a long straight vertical wire and a solenoid. The directions of the currents are indicated by arrows. Add to the diagrams the magnetic fields generated by each of these conductors alone.

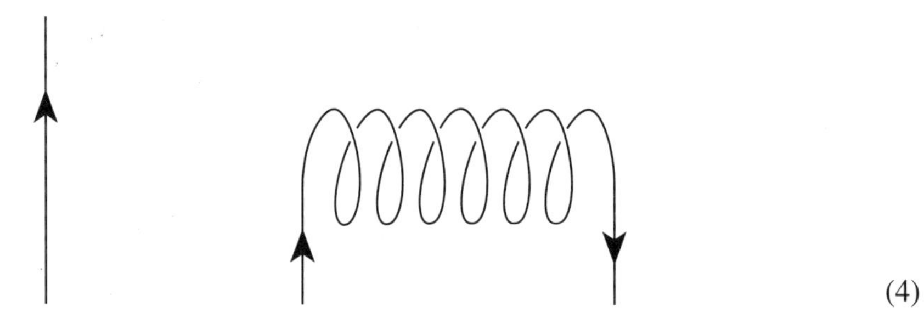

(4)

(b) The diagram below shows two long straight parallel conductors, K and L. P is a point midway between the wires and PQ is a line perpendicular to the plane containing the wires. The direction of the current through wire K is shown. The resultant field at P is zero.

On the diagram draw arrows to show the following.

(i) The direction of the current through wire L.

(ii) The direction of the force acting on wire K due to the current through wire L.

(iii) The direction of the force acting on wire L due to the current through wire K. (3)

(c) What is the direction of the resultant field along the line from P towards Q?

....

Sketch a graph to show how you would expect the magnitude of this field to vary with distance along PQ.

(4)

ULEAC

6 Blood contains ions in solution. The diagram shows a model used to demonstrate the principle of an electromagnetic flowmeter which is used to measure the rate of flow of blood through an artery.

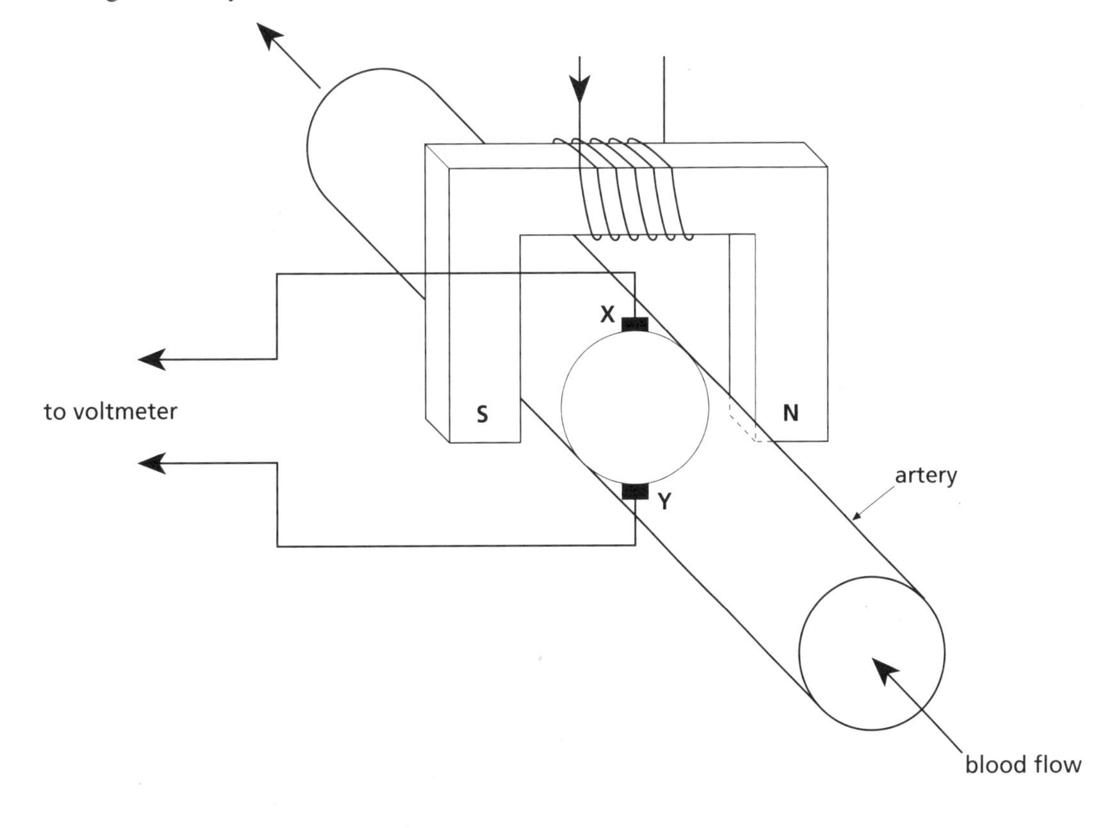

QUESTIONS

When a magnetic field of 2.0 T is produced by the electromagnet, a potential difference (p.d.) of 600 μV is developed between the two electrodes **X** and **Y**. The cross-sectional area of the artery is 1.5×10^{-6} m^2. The separation of the electrodes is 1.4×10^{-3} m.

(a) Write down an expression for the force on an ion in the blood which is moving at right angles to the **magnetic** field, defining the symbols you use.

....

....

(b) An ion has a charge of 1.6×10^{-19} C.
Show that the force on an ion due to the **electric** field between **X** and **Y** is 6.9×10^{-20} N.

....

....

(c) Given that a p.d. of 600 μV is developed when the electric and magnetic forces on an ion are equal and opposite, calculate:

(i) the speed of the blood through the artery;

....

....

(ii) the volume of blood flowing each second through the artery.

....

....

.... (8)

AEB

7 (a) A long bar magnet hangs from one end of a spring, as shown in the diagram.

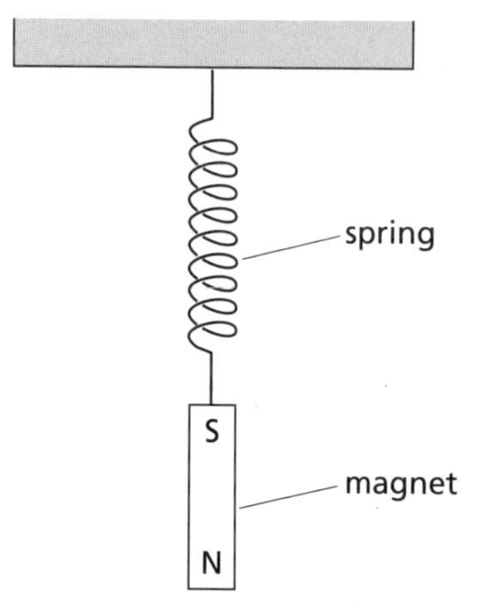

The magnet is displaced vertically downwards and then released. The subsequent vertical displacement x is found to vary with time t as shown in the diagram below.

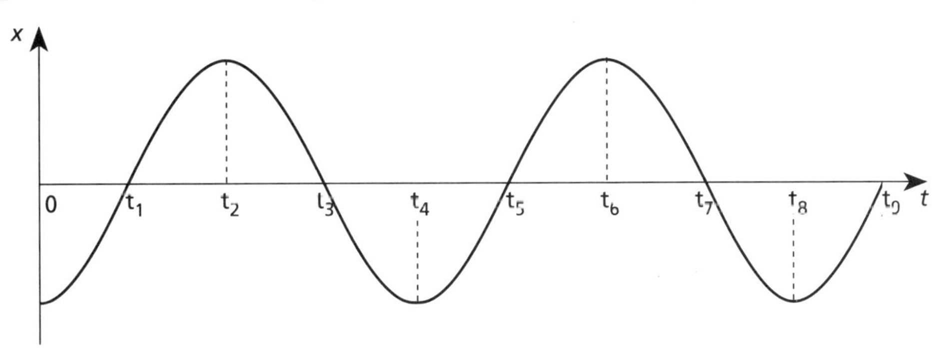

(i) State **two** times, apart from $t = 0$, at which the magnet is stationary.

....

(ii) State **two** times at which the magnet is moving vertically upwards with maximum speed.

....

(iii) State **two** times at which the magnet is moving vertically downwards with maximum speed.

.... (3)

(b) The north pole of the magnet is now placed inside a coil of wire, as shown in the diagram below.

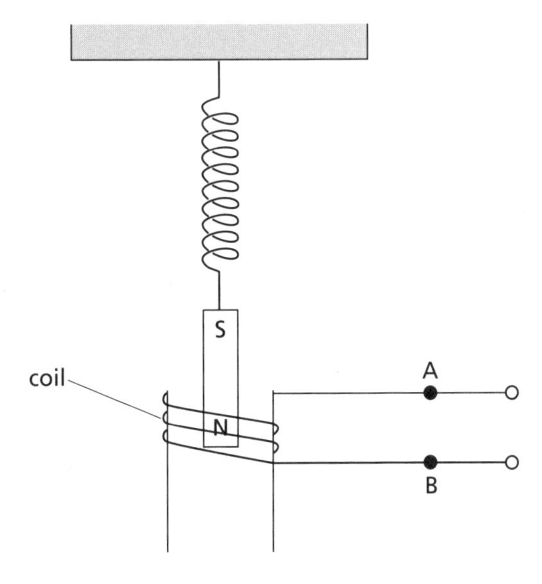

The terminals of the coil are connected to the Y-plates of a cathode-ray oscilloscope (c.r.o.) which may be assumed to have infinite input resistance.

QUESTIONS

(i) Sketch a graph to show how the induced e.m.f. in the coil will vary with time t when the magnet oscillates in the coil. Mark relevant times (for example, t_1, t_2, t_3) on the t-axis of your graph.

(ii) Use the laws of electromagnetic induction to explain the shape of your graph.

....

....

....

.... (7)

(c) A high resistance resistor is now connected in parallel with the c.r.o. between the points A and B (see the diagram on p. 57).

(i) Draw a second graph to show how the e.m.f. will vary with time t.

(ii) Explain, in terms of the principle of conservation of energy, why this graph is different from your first graph.

....

....

....

....

(iii) Describe, with the aid of a sketch graph, the changes which would occur in the shape of the graph drawn in (c)(i) if the resistance of the resistor has been reduced to a very low value.

....

....

.... (10)

UCLES

8 (This question is based on experiments conducted by NASA on a novel energy source for a space shuttle.)
Above the equator of the Earth, the Earth's magnetic field is horizontal and runs south-north. A space shuttle is to travel west-east across the magnetic field, as shown in the diagram. A small satellite is attached to the shuttle by a very long thin copper cable. It is proposed to use the e.m.f. generated in this cable as a source of electrical energy for the shuttle.

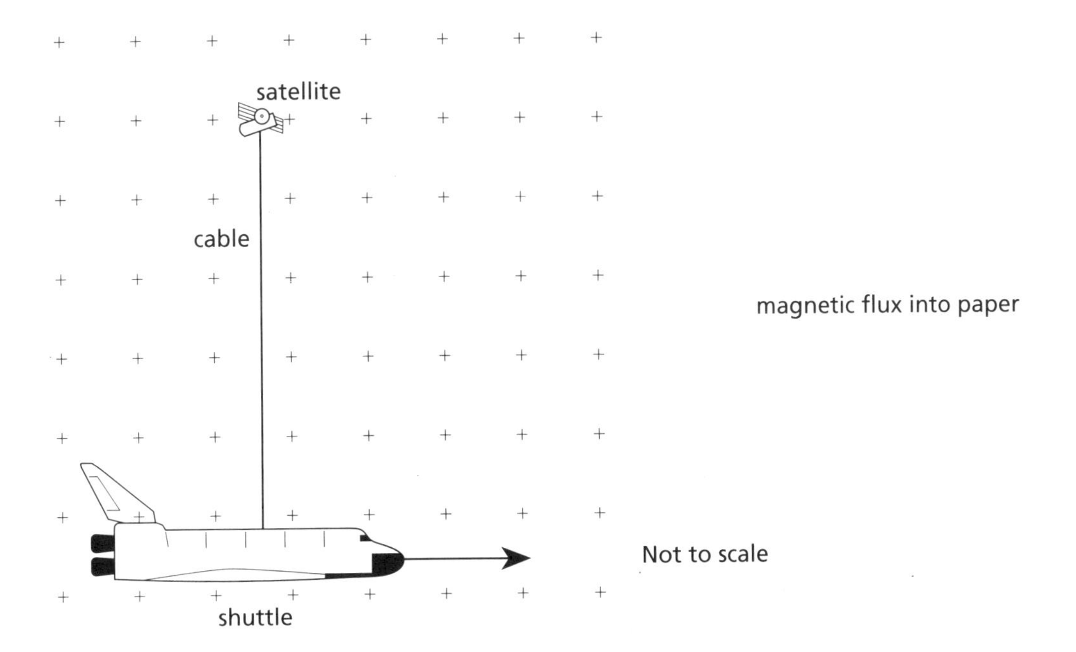

(a) (i) Explain why an e.m.f. will be generated in the cable.

....

.... (2)

(ii) Calculate the magnitude of this e.m.f. using the following data:
Cable length = 20 km
Speed of shuttle and satellite = 7.0 km s^{-1}
magnetic flux density of the Earth's field = 60×10^{-6} T. (3)

....

....

(b) It has been suggested that this e.m.f. could be used to power equipment on board the shuttle if two cables were used side-by-side so that there was a complete circuit. Explain why this arrangement would not work.

....

....

.... (2)

Oxford

Conduction, **convection**, and **radiation** are three ways in which thermal energy can be transferred within and between objects.

All objects emit electromagnetic **radiation**. As the temperature of an object is increased, the power it radiates also increases, as does the frequency of the highest-energy photons. Natural **convection currents** occur in fluids and are due to the change in density when part of the fluid is heated or cooled. There is a net upward force on material that is less dense than its surroundings and a net downward force on material that is denser than the surroundings. **Forced convection** describes the movement of a fluid such as the air in a hairdryer which is forced over the heating element by a motor.

The mechanism of **conduction** involves the transfer of energy from molecule to adjacent molecules through intermolecular forces. In insulating solids, liquids and gases this is a slow process. Conduction in metals is a much more rapid process because the energy is also, and mainly, transferred by the diffusion of free electrons.

The rate at which thermal energy is conducted depends on the **thermal conductivity** of the material, its cross-sectional area and the **temperature gradient**. This is described by the equation

$$\frac{Q}{t} = kA\frac{\Delta\theta}{l}.$$

The U-value of a material or combination of materials describes the rate of energy transfer in terms of the **temperature difference**,

$$\frac{Q}{t} = UA\Delta\theta .$$

Temperature is a measure of the average energy of the atoms and molecules of an object; the total amount of this energy is called its **internal energy**. A change in the internal energy can be brought about by **heating** the object or **doing work** on it. The **first law of thermodynamics** states that the *increase in internal energy (ΔU) of an object is equal to the sum of the thermal energy transferred to it (ΔQ) and the work done on it (ΔW)*. The phrases in *italics* are definitions of the terms in the formula $\Delta U = \Delta Q + \Delta W$.

There is also a **zeroth law of thermodynamics**: if two objects are in thermal equilibrium with a third then they are in thermal equilibrium with each other; this defines the meaning of 'two things are at the same temperature' as being 'if they were placed together there would be no net energy transfer between them'.

Adiabatic changes are those where there is no thermal energy transfer into or out of a system, hence $\Delta U = \Delta W$. **Isothermal** changes are at constant temperature, so the internal energy does not change, $\Delta Q = -\Delta W$. When the volume of a gas is changed at constant pressure the work done is equal to the area between the graph line and the volume axis of a p-V graph; this is equivalent to the formula $W = p\Delta V$. The **ideal gas equation**

$$\frac{p_1V_1}{T_1} = \frac{p_2V_2}{T_2}$$

applies to any change that occurs to a fixed mass of gas.

The term **latent heat** refers to the energy required to change the state of a substance e.g. solid to liquid, without a change in temperature. **Specific latent heat**, L, is the energy transfer per kilogram;

$$L = \frac{Q}{m}.$$

Molar latent heat, L_m, is the energy transfer per mole;

$$L_m = \frac{Q}{n},$$

where n represents the number of moles of substance. **Heat capacity** is the heat energy required to raise the temperature of an object by 1°C. When applied to a substance there are two

REVISION SUMMARY

alternative ways of measuring it; either **specific heat capacity**, the energy per kilogram $Q = mc\Delta\theta$ or **molar heat capacity**, the energy per mole $Q = nc_m\Delta\theta$. Different values are obtained when the heat capacity of a gas is measured at constant pressure and at constant volume. The molar heat capacity of a gas measured at constant pressure, c_p, is greater than that measured at constant volume, c_v, because of the extra work that has to be done to push away the surrounding air when the gas expands on heating under constant pressure.

Thermodynamic temperature, which is the absolute scale of temperature, is defined using the form of the **ideal gas equation**, $pV = nRT$. This fixes **absolute zero** at 0 K and the kelvin as 1/273.16 of the temperature of the triple point of water. Celsius temperature is defined as being the kelvin temperature – 273.15 (hence 0°C is 273.15 K).

While the Celsius and kelvin scales assign a unique value to any temperature, measurements of the same temperature made on a centigrade scale may vary according to the thermometric property used. A centigrade scale of temperature is set up by taking measurements of the property, e.g. resistance or length of a mercury column, at the ice point (X_0) and the steam point (X_{100}) and dividing the interval into one hundred equal units. A temperature θ can then be determined by measuring the property at that temperature (X_θ) and using the formula

$$\frac{\theta}{100} = \frac{X_\theta - X_0}{X_{100} - X_0}.$$

If you need to revise this subject more thoroughly, see the relevant topics in the Letts A level *Physics Study Guide*.

QUESTIONS

1 1 kg of ice at 0°C is placed in an insulated container of negligible heat capacity and heated with a constant power supply. During 1 hour, the ice melts, the water rises in temperature to 100°C and is then entirely vaporised.
specific heat capacity of water = 4 kJ kg^{-1} K^{-1}
specific latent heat of fusion of ice = 3×10^2 kJ kg^{-1}
specific latent heat of vaporisation of water = 2×10^3 kJ kg^{-1}
At approximately what time after starting is the temperature of the water 50°C?

A 6 minutes
B 12 minutes
C 24 minutes
D 30 minutes (1)

Oxford

2 A length of platinum wire is used as a resistance thermometer. The wire has a resistance of 20.0 Ω at the ice-point and 27.8 Ω at the steam-point. The wire is then placed in a bath of hot oil, and the resistance is found to be 39.5 Ω. What is the temperature of the oil, as measured on the centigrade scale of this platinum resistance thermometer?

A 150°C
B 250°C
C 282°C
D 350°C
E 506°C (1)

NICCEA

3 A fixed mass of an ideal gas initially has a volume V and an absolute temperature T.
Its initial pressure could be doubled by changing its volume and temperature to
A $V/2$ and $4T$
B $V/4$ and $T/2$
C $2V$ and $T/4$
D $4V$ and $2T$ (1)

AEB

4 Electrical conduction can be compared to thermal conduction. Which electrical quantity is analogous to *the rate of energy transfer*?

A current
B potential difference
C power
D charge (1)

5 A block of metal, of mass 103 g, is heated to 100°C and then transferred to a polystyrene beaker containing 200 g of water at 19.8°C. When thermal equilibrium has been reached, the water temperature is 21.6°C.
Calculate the energy gained by the water during this process and the specific heat capacity of the metal.
(The specific heat capacity of water is 4.20 kJ kg^{-1} K^{-1})

....

....

....

....

....

The experiment as described is not a particularly good one for measuring the specific heat capacity of the metal. Name *two* important sources of error. State a way of reducing *one* of them.

....

....

.... (9)

ULEAC

6 A pupil uses the apparatus below to investigate the properties of a sample of gas.

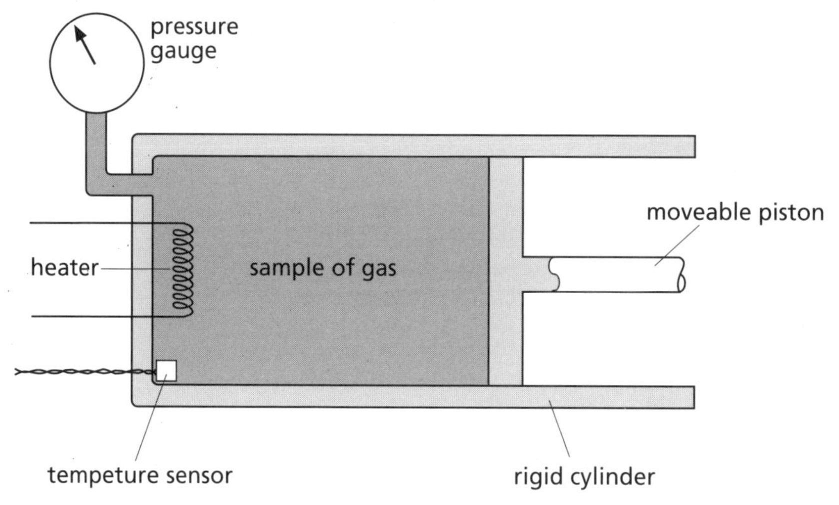

The volume of the sample of gas can be changed by moving the piston.
The temperature of the sample of gas can be increased using the heater.
At the start, the pressure of the gas is 400 kPa and its volume is 1000 cm^3.
During the investigation, the pressure and volume of the gas change as indicated by sections AB and BC on the graph below.

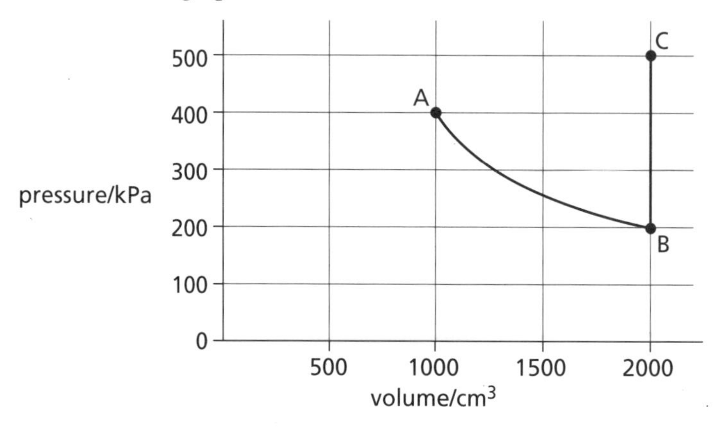

During section AB, the temperature of the gas is constant at 300 K.

(a) Calculate the volume of the gas when its pressure is 250 kPa during stage AB.

....

.... (2)

(b) State what happens to the pressure, volume and temperature of the gas over the section of the graph which starts at B and finishes at C.

....

.... (2)

(c) What is the temperature of the gas, in kelvin, corresponding to point C on the graph?

....

.... (2)

SEB

7 The diagram shows a typical arrangement for a domestic hot-water tank. The water can be heated by an immersion heater and the tank has lagging around the walls and over the top.

outlet
lagging
copper tank
immersion heater
A
B
C
vertical cross-section

(a) Why is convection important in heating the water?

....

.... (1)

(b) Why is the bottom of the tank not lagged?

....

....

....

.... (2)

QUESTIONS

(c) Sketch a graph on the axes below to show how the temperature is likely to vary along the line of points ABC shown in the diagram above. Assume the tank surface is flat in this region.

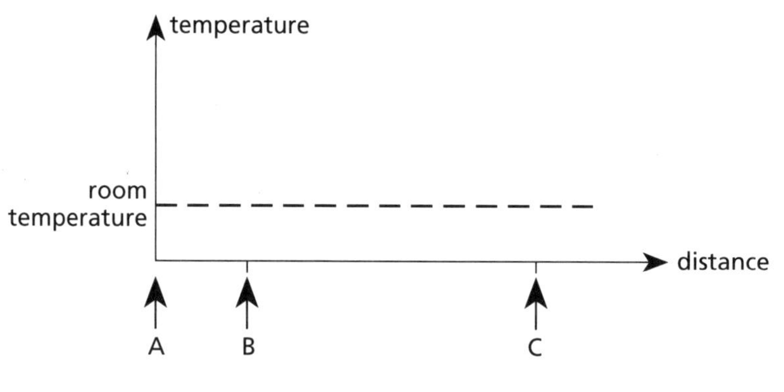

(3)
UCLES

8 (a) The distance between the 0°C mark and the 100°C mark on the stem of a mercury-in-glass thermometer is 18.0 cm. How far is the 35°C mark from the 100°C mark?

....

....

....

Distance from 100°C mark = ______________ (3)

(b) Calculate the energy transferred when 200 g of ice at 0°C melts and warms up to 35°C.
Specific latent heat of fusion of ice = 3.4×10^5 J kg^{-1}
Specific heat capacity of water = 4200 J kg^{-1}K^{-1}

....

....

....

Energy transferred = ______________ (3)
ULEAC

9 (a) Conditions in the Earth's atmosphere are largely affected by rising air currents. Air, heated at the Earth's surface, rises. As it continues to rise its pressure and temperature fall as the air undergoes adiabatic expansion.

(i) Explain briefly why heated air rises.

....

....

.... (3)

(ii) What is meant by the term *adiabatic*?

.... (1)

(iii) Apply the first law of thermodynamics to show that an adiabatic expansion must be accompanied by a fall in temperature.

....

....

....

....

....

.... (4)

(b) An airline pilot, ascending from take-off, notes that at a height of 2500 m the temperature and atmospheric pressure have fallen to 273 K and 75 kPa. The temperature and pressure on the ground were 287 K and 101 kPa respectively.
[The gas constant $R = 8.3\ \text{J mol}^{-1}\,\text{K}^{-1}$. The molar mass of air $M = 0.029\ \text{kg mol}^{-1}$.]

(i) Calculate the density ρ of the air at the mean temperature and the mean pressure during the ascent;

....

....

....

.... (3)

(ii) Calculate the pressure difference between top and bottom of an isolated 2500 m static vertical air column. Assume the mean density of the air is the value calculated in (i). Take the acceleration of free fall g in this calculation as $9.8\ \text{m s}^{-2}$.

....

....

....

.... (3)

(iii) If the aircraft is ascending through low cloud, what flight hazard might be expected on reaching the height of 2500 m? Give a reason for your answer.

....

....

.... (2)

(c) The atmospheric pressure at a cruising height of 10 000 m is 26 kPa, but the pressure within the airliner is maintained at 95 kPa. The windows by the passenger seats each have an area of 0.12 m^2 and each is secured by 20 bolts.
Calculate the mean tension in the bolts caused by the pressure difference between atmospheric pressure and that in the passenger spaces.

....

....

....

....

....

.... (4)

Oxford

10 (a) Explain how a physical property of a substance which varies with temperature may be used for the measurement of temperature.

....

....

.... (2)

(b) (i) Describe the principal features of one type of liquid-in-glass thermometer.

....

....

....

(ii) Discuss the relative advantages and disadvantages of a liquid-in-glass thermometer and a resistance thermometer which may be used in the same temperature range.

....

....

....

....

.... (7)

(c) A resistance thermometer is placed in a bath of liquid at 0°C and its resistance is found to be 3740 Ω. At 100°C, its resistance is 210 Ω. The bath is now cooled until the resistance of the thermometer is 940 Ω.

(i) What is the temperature of the bath, as measured using the resistance thermometer?

....

....

....

(ii) The reading taken at the same time on a mercury-in-glass thermometer placed in the bath is 40°C. Suggest a reason for the difference between this reading and the value calculated in (c)(i).

....

.... (3)

(d) (i) What do you understand by the absolute (thermodynamic) scale of temperature?

....

....

(ii) The pressure p of an ideal gas of density ρ is related to the mean square speed $<c^2>$ of its molecules by the expression

$$p = \frac{1}{3}\rho<c^2> .$$

QUESTIONS

Deduce an expression for the thermodynamic temperature T of the gas in terms of the mean kinetic energy $<E_k>$ of a molecule at that temperature.*

....

.... (5)

(e) Explain, in terms of the energies of atoms, conditions under which it is possible to increase the total energy of the atoms of a substance without any change of temperature of that substance.

....

....

.... (3)

* this part of the question relies on material from Unit 8 *UCLES*

REVISION SUMMARY

Hooke's law states that the extension of a material is proportional to the tension. This law does not apply to all materials, and when it does apply it is only over a limited range of tensions. The **stiffness** of a material is measured by the **Young modulus**,

$$E = \frac{\text{stress}}{\text{strain}},$$

where stress is the force per unit cross-sectional area,

$$\sigma = \frac{F}{l},$$

and strain is the fractional increase in length,

$$\varepsilon = \frac{e}{l}.$$

Combining these gives the formula for the Young modulus

$$E = \frac{Fl}{eA}.$$

The energy stored in a stretched object such as steel wire is called **elastic strain energy** and is given by the formula $E = {}^1/_2 Fe = {}^1/_2 \times \sigma \times \varepsilon \times V$, where V is the volume of the material. Note that in this paragraph E has been used to represent two different physical quantities.

A sample of material shows **elastic** behaviour if it returns to its original shape when the deforming force is removed; beyond the **elastic limit** it is said to be **plastic**. Metal samples, when stretched beyond the elastic limit, reach a point where they 'give', i.e. they become easier to stretch. When this happens the stress is called the **yield stress**. The **ultimate tensile stress** is the stress required to break the material.

Solids, liquids and gases have different structures. The evidence for gases being composed of widely-spaced molecules in rapid, random motion comes from **Brownian motion**. This is the motion of large particles such as smoke specks suspended in air. The smoke specks show a disordered, jerky motion that can be explained in terms of their continual bombardment by air molecules moving at high speeds in all directions. **Diffusion** gives further evidence of molecular motion; in a gas it is a rapid process, indicating that molecules can move relatively large distances between collisions. Diffusion in a liquid is slow, providing evidence that the molecules are in close contact and have very little freedom of movement.

Crystalline solids such as metals have their atoms arranged in a regular lattice within each crystal; the crystals are arranged in a random manner. **Polymeric** materials such as nylon, rubber and PVC are composed of long chain molecules. Glass has an **amorphous** structure, there is no pattern to the molecular arrangement, it is totally disordered.

The **kinetic theory** pictures a gas consisting of atoms or molecules in constant motion, undergoing elastic collisions with each other and the walls of the containing vessel. **Pressure** is due to the force exerted when gas molecules collide with objects and walls. By making simplifying assumptions it can be shown that the pressure exerted by a gas is given by the formula $p = {}^1/_3 nm \langle c^2 \rangle$, where n represents the number of molecules per unit volume, m the mass of each molecule and $\langle c^2 \rangle$ is the **mean square speed** of the molecules. It follows from this that temperature is proportional to the mean molecular kinetic energy. **Boltzmann's constant**, k, relates the mean molecular kinetic energy to the temperature in the formula for a monatomic gas $E_k = {}^3/_2 kT$. The Avogadro constant, N_A, the number of entities in a mole, is related to the gas constant R and the Boltzmann constant k by the formula

$$k = \frac{R}{N_A}.$$

Evidence for atoms having a dense, positively charged nucleus comes from the **scattering of alpha particles**. Extra-nuclear **electrons** orbit the nucleus; **ionisation** occurs by adding or taking electrons away from those in orbit. The existence of **line spectra** provides evidence that the electrons are restricted to orbits that correspond to certain **energy levels**. Each line in the

REVISION SUMMARY

spectrum corresponds to an electron transition between two of these levels at energy levels E_1 and E_2, the frequency of the light emitted being given by the formula $hf = E_1 - E_2$.

The structure of an atom can be represented by the symbol $^A_Z X$. A represents the **mass number**, the total number of neutrons and protons, collectively called nucleons, in the nucleus. Z is the atomic number, the number of protons in the nucleus. **Isotopes** of an element have the same atomic number but a different mass number due to having different numbers of neutrons. A nucleus has less potential energy than its constituent protons and neutrons; work would have to be done against the nuclear force to separate the nuclear components. This difference in energy is called the **nuclear binding energy**; it has a negative value. Einstein showed that energy has mass given by the equation $E = mc^2$; since a nucleus has less energy than its components it also has less mass. Because the **binding energy per nucleon** becomes numerically greater with increasing mass number up to a mass number of about 50, and then becomes numerically smaller, energy can be released by **fusion** of small atoms to make larger ones and **fission** of larger atoms to make smaller ones. The nuclear reactions in stars are fusion reactions; those in Earth-based reactors are **fission** ones.

Radioactive decay is the random process by which an unstable nucleus undergoes changes that result in it becoming more stable. (Mass is lost in the process, the mass lost appearing as energy.) Any one or a combination of **alpha particles**, **beta particles** and **gamma radiation** can be emitted when this happens. The table summarises the properties of these emissions.

type of nuclear radiation	nature	charge	penetration	ionising ability
alpha	two neutrons and two protons, sometimes referred to as a helium nucleus	positive $+2e$	stopped by a few cm of air or a thin piece of card	intensely ionising
beta	high-speed electron emitted when a neutron decays to a proton and an electron	negative $-e$	partially absorbed by aluminium foil; totally absorbed by 5 mm aluminium	less than alpha; ionisation occurs at collisions with atoms and molecules
gamma	high frequency, short wavelength electromagnetic radiation	none	never totally absorbed; intensity is reduced by thick lead or concrete	weakly ionising; the high penetration is due to few collisions where ionisation would occur

If you need to revise this subject more thoroughly, see the relevant topics in the *Letts A level Physics Study Guide.*

Although the decay of individual nuclei is random, statistically the rate of decay is proportional to the number of undecayed nuclei present in a sample. The **decay constant of an isotope**, λ, relates these quantities in the formula

$$\frac{dN}{dt} = -\lambda N .$$

The solution of this differential equation, $N = N_0 e^{-\lambda t}$, can be used to calculate the number of undecayed nuclei (N) at time t after a known number, N_0, were present.

The **half-life**, $t_{1/2}$, the average time taken for the rate of decay to halve (or the time for half the radioactive nuclei to decay), is also a constant for a particular isotope and is related to the decay constant by the formula $\lambda t_{1/2} = \ln 2$.

QUESTIONS

1 Which of the following statements is **not** consistent with the kinetic theory of an ideal gas?

A The molecules occupy a small fraction of the volume of the container.
B The molecules collide inelastically.
C The average kinetic energy of the molecules gives a measure of the temperature of the gas.
D The attractive forces between the molecules are negligible. (1)

SEB

2

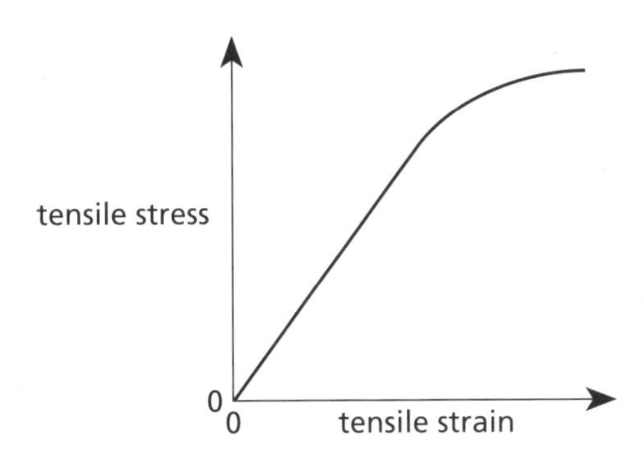

The diagram shows a stress-strain graph for a specimen of a material.
Which statement(s) about the graph is/are correct?

1 The graph shows the behaviour of a typical brittle material.
2 For small strains the material obeys Hooke's law
3 The gradient of the graph below the limit of proportionality gives a value for the Young's modulus.

A 1 and 2 only correct
B 2 and 3 only correct
C 1 only correct
D 3 only correct (1)

AEB

3 Two identical wire cables support a concrete block of mass 1000 kg as shown in the diagram. The load causes both cables to extend by 1 mm. Free-fall acceleration may be taken as 10 m s^{-2}.

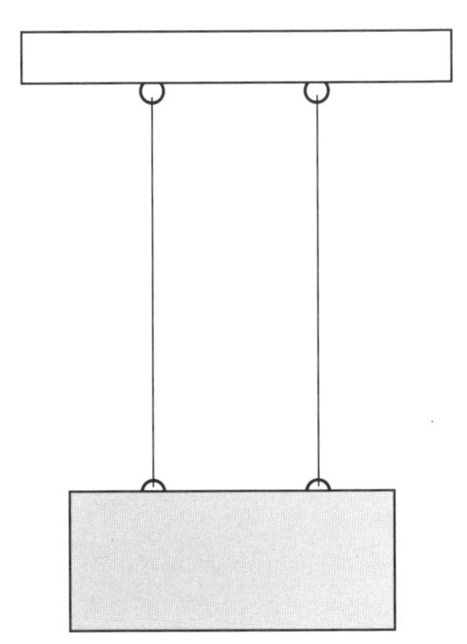

QUESTIONS

Applying the Hooke's law expression *Force* = $k \times$ *extension*
the value of the constant k for **each** of these cables is

A 5×10^5 N m^{-1}
B 1×10^6 N m^{-1}
C 5×10^6 N m^{-1}
D 1×10^7 N m^{-1}

(1)
Oxford

4 A radioactive isotope has a half-life of 20 minutes. A particular sample of this isotope gives a count rate of 3200 per second at 2 o'clock on a certain afternoon.
At what time on that day is the count rate 200 per second?

A 3.00 p.m.
B 3.20 p.m.
C 3.40 p.m.
D 4.40 p.m.
E 7.20 p.m.

(1)
SEB

5 Calculate the binding energy, in joules, of the oxygen nuclide $^{18}_{8}O$.

Particle	Mass
proton	1.008 u
neutron	1.009 u
$^{18}_{8}O$	17.999 u

1 u (atomic mass unit) $= 1.660 \times 10^{-27}$ kg

....

....

....

.... (4)
SEB

6 (a) The three diagrams represent stress-strain graphs for different materials up to the point at which they break.

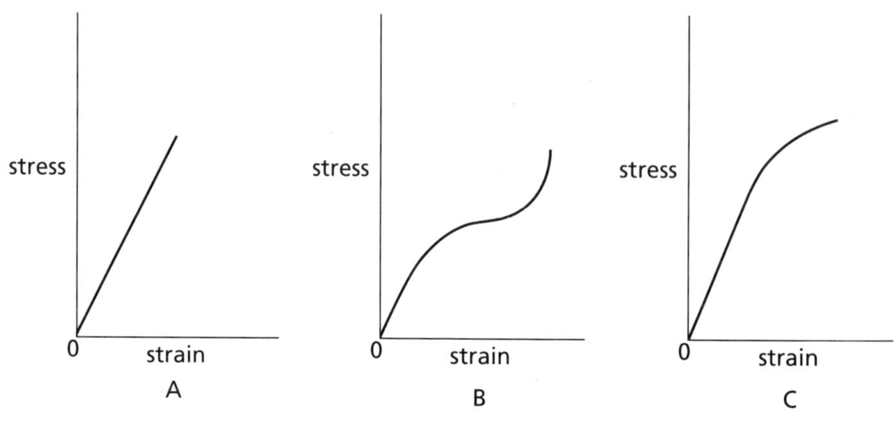

QUESTIONS

Which of these graphs best represents the behaviour of the following:

(i) a thin glass fibre; __________ (1)
(ii) a copper wire; __________ (1)
(iii) a rubber band? __________ (1)

(b) Candle-wax, diamond and nylon are materials with different structures. Which of these materials is:

(i) amorphous; __________ (1)
(ii) polymeric; __________ (1)
(iii) crystalline? __________ (1)

(c) (i) Explain what is meant by *plastic behaviour*.

....

....

.... (2)

(ii) State which of the materials referred to in (b) exhibit plastic behaviour.

....

.... (2)

Oxford

7 (a) The diagram shows a copper wire of cross-sectional area A and unstretched length L_0.

When one end of the wire is clamped and a force F applied to the other, the wire is stretched to a length L, as shown in the diagram below (not to scale).

(i) Write down expressions for the stress σ and strain ε in the wire in terms of A, F, L_0 and L.

Stress σ = __________
Strain ε = __________ (2)

The stress is gradually increased. The graph overleaf shows how the strain of the wire depends on the stress.

QUESTIONS

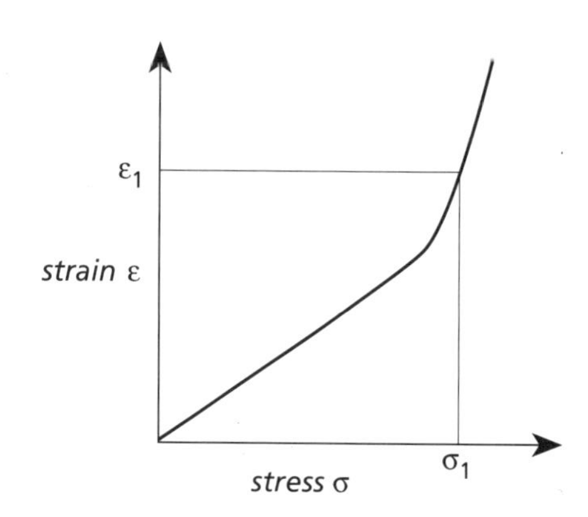

(ii) State how a value of the Young's modulus for copper could be obtained from this graph.

....

.... (2)

(iii) State how the work done in giving the wire the strain ε_1 could be obtained from the graph, together with any other data required.

....

.... (2)

(b) A steel wire is fixed horizontally between two clamps P and Q, 2000 mm apart, so that it is taut but not under tension. The cross-sectional area of the wire is 7.8×10^{-7} m^2. When a certain load W is attached to the mid-point of the wire, it is found that the equilibrium position of the point of attachment is 100 mm below PQ. The equilibrium tension in the wire is T. The forces acting are shown in the diagram below (not to scale).

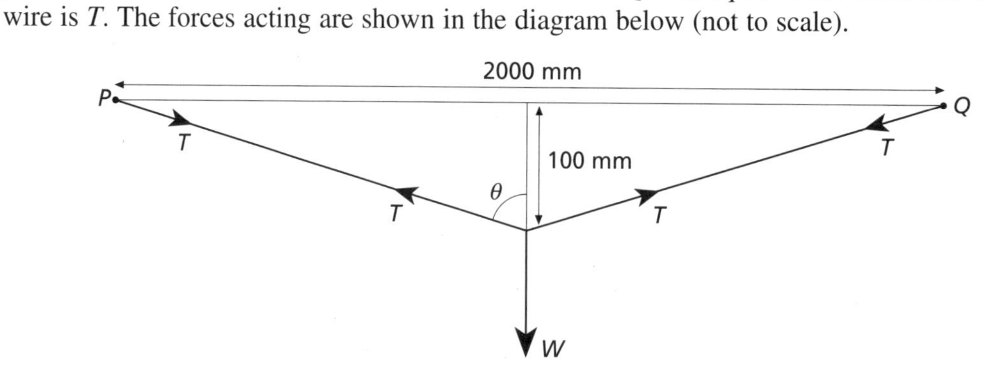

(i) Calculate the length L_1 of the wire when it is in the equilibrium position shown in the diagram above. (Give your answer to the nearest mm.)

....

....

L_1 = ________ mm (2)

QUESTIONS

(ii) Using the value 2.1×10^{11} Pa for the Young's modulus of steel, deduce the tension T in the wire when it is in this position.

....

....

$T =$ ________ N (3)

(iii) Write down the equation relating W to T and the angle θ when the load is in equilibrium.

.... (1)

(iv) Calculate the angle θ when the load is in equilibrium.

....

....

$\theta =$ ________ ° (2)

(v) Use your answers to (b) (ii), (iii) and (iv) to calculate the value of W.

....

$W =$ ________ N (1)

NICCEA

8 (a) Gas pressure is caused by molecular bombardment. Explain in molecular terms why the pressure of a gas increases when its temperature is increased at constant volume.

....

....

....

(b) The radiation from some stars can ionise the surrounding interstellar gas causing it to glow. Astronomers call the outcome an emission nebula.
Interstellar gas is mostly monatomic hydrogen at a temperature of 1.0×10^4 K and a concentration of 1.0×10^7 atoms per cubic metre.

Mass of a hydrogen atom = 1.7×10^{-27} kg
Molar mass of monatomic hydrogen = 1 g
Molar gas constant, R = 8.3 J $K^{-1}mol^{-1}$

QUESTIONS

Determine:

(i) the mass of one cubic metre of interstellar gas;

....

....

(ii) the pressure of the interstellar gas;

....

....

(iii) the root-mean-square speed of the atoms

....

.... (8)

AEB

9 A certain radioactive source emits only gamma radiation.
A technician is asked to determine the half-value thickness of lead for the radiation from this source.
The technician sets up the apparatus shown below and keeps the distance between the source and the gamma ray detector the same throughout the experiment.

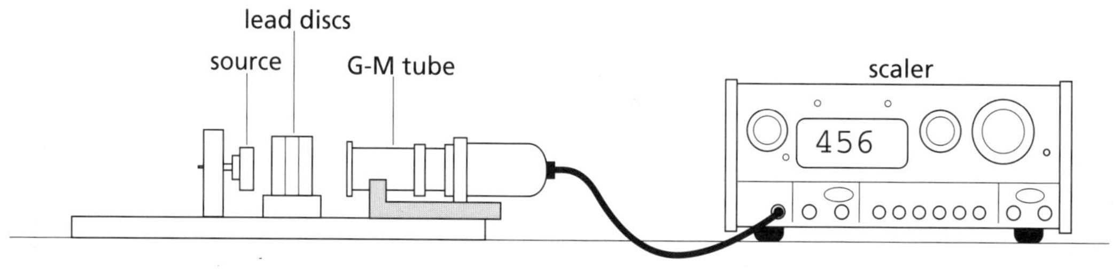

The technician measures the count rate several times for a certain thickness of lead sheet and obtains an average value for the count rate.
The measurements are repeated with several different thicknesses of lead sheet and also with no lead present.
The source and the lead are then removed and the background count rate is measured.
The technician corrects each average count rate for background and records the results as shown in the table.

Thickness of lead sheet in mm	Corrected average count rate in counts/minute
0	520
5	390
10	280
15	200
20	145
25	110

QUESTIONS

(a) (i) Draw a graph of corrected average count rate against thickness of lead sheet, using square-ruled paper.
Find the half-value thickness of lead for this source.

....

(ii) On the same axes, sketch a graph which might be obtained if the average count rate was not corrected for background radiation. (5)

(b) 21 years later, another technician repeats the experiment with the same source.
The gamma ray source has a half-life of 5.25 years.
What corrected average count rate would be recorded with no lead sheet between the source and the detector?

....

....

.... (2)

SEB

10 (a) What is meant by

(i) the *decay constant* λ of a radioactive material,

....

....

(ii) the *half-life* $t_{1/2}$?

....

.... (2)

(b) The decay constant and half-life are related by the equation:

$$\lambda = \frac{0.693}{t_{1/2}}$$

The half-life of $^{60}_{27}\text{Co}$ is 5.26 years.

(i) What do the numbers 27 and 60 represent?

27 ....

60 .... (2)

QUESTIONS

(ii) Calculate the decay constant of $^{60}_{27}Co$.

(1)

(iii) Calculate the activity of 1.00 gram of $^{60}_{27}Co$

(60 grams of $^{60}_{27}Co$ contain 6.02×10^{23} atoms.)

(3)
UCLES

QUESTIONS

1 An experiment was performed to investigate the absorption of γ-rays by lead using a cobalt-60 source and a Geiger-Müller tube connected to a counter. A graph of N, the number of counts recorded in one minute corrected for background, against x, the thickness of absorber, was plotted.

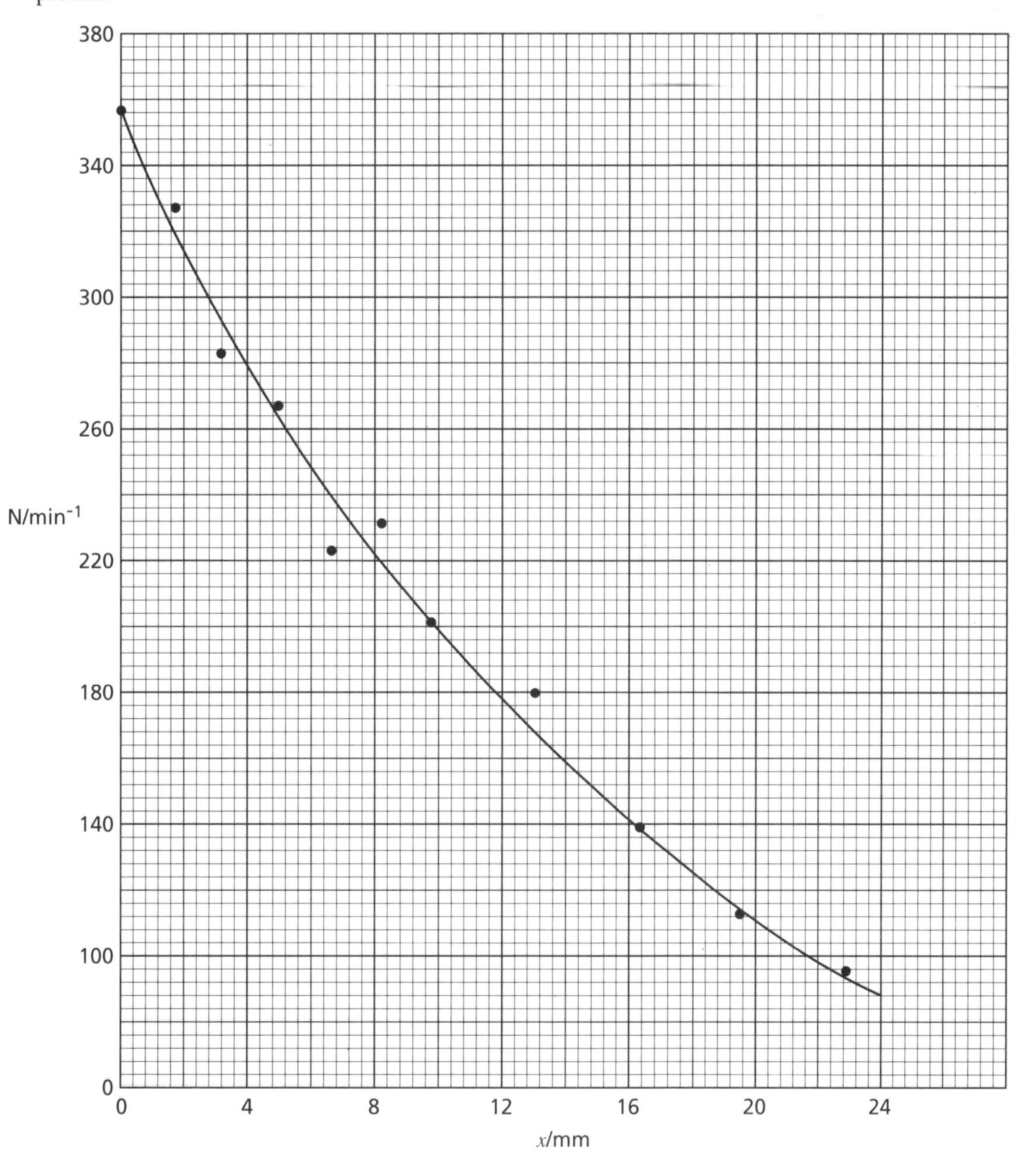

(i) Explain the likely cause of the 'scatter' in the points and suggest how this could have been reduced.

....

....

....

.... (3)

QUESTIONS

(ii) In such an experiment the 'stopping ability' t of the absorber is often expressed in the form

$$t = \rho x$$

where ρ is the density of the absorbing material and x is the thickness of the absorber. The density was found to be 1.06×10^4 kg m^{-3} by a subsequent experiment. Use the graph to complete the table below.

x/mm	N/min^{-1}	ln(N/min^{-1})	t/kg m^{-2}
0.00			
2.00			
4.00			
6.00			
8.00			
10.00			
12.00			
14.00			

(4)

(iii) Plot a graph of ln(N/min^{-1}) against t. (5)

(iv) Determine a value for the mass absorption coefficient μ for the radiation, given that $-\mu$ is equal to the gradient of this graph.

....

....

....

.... (4)

If you need to know more about data analysis, see Chapter 9 of the *Letts A level Physics Study Guide.*

(v) A radiation data book gives the following information for γ-rays.

Energy/MeV	μ/10^{-3} m^2 kg^{-1}
0.80	8.75
1.00	7.04
1.20	6.13
1.40	5.42
10.00	5.01

Use this data to estimate a value for the energy of γ-rays from cobalt-60, explaining how you arrived at your answer.

....

....

.... (2)

ULEAC

1 FORCES AT REST

Question	Answer	Mark
1	D	1

Examiner's tip Acceleration, momentum, force and displacement are all vector quantities. The others are scalars, they do not have a direction.

Question	Answer	Mark
2	C	1

Examiner's tip **Y** is equal in size (and opposite in direction) to the horizontal component of **X**. The vertical component of **X** is equal and opposite to **Z**.

Question	Answer	Mark
3	A	1

Examiner's tip The speed of 3 $m s^{-1}$ is irrelevant. You should recognise that, as the cyclist is travelling at constant velocity, the forces are balanced. The resistive forces on the cyclist act in the direction opposite to the motion. The diagram shows the forces on the cyclist.

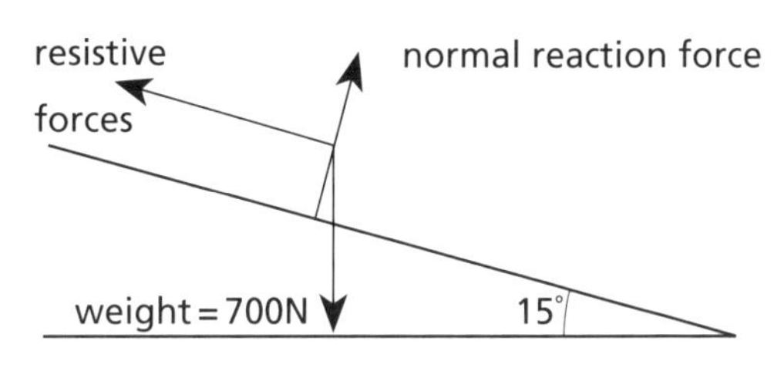

The component of the weight perpendicular to the slope (700 cos 15°) balances the normal reaction force and the component parallel to the slope (700 cos 75°) balances the resistive forces.

Question	Answer	Mark
4	D	1

Examiner's tip The downward force at **X** prevents rotation about **Y**. Using the principle of moments, this force must equal 900 N (from 600 N × 5 m = F_Y × 2 m). The force at **Y** is calculated using the fact that, for equilibrium, the total upward force must equal the total downward force (F_Y = 600 N + 900 N).

Question	Answer	Mark
5	$T \cos 20° = 500$ N	1
	$T = 500 \div \cos 20° = 532$ N	1

Examiner's tip The principle is that the horizontal component of the tension must be equal in size to the frictional force which is acting in the opposite direction. The vertical component of the tension opposes the weight of the log.

Question	Answer	Mark
6 (a)	The person is travelling at a constant velocity.	**1**

Examiner's tip It is important to state that both the speed and the direction of travel are not changing. The term velocity describes both speed and direction.

Question	Answer	Mark
(b)	The vector diagram is:	**1**

D
0.65 kN
50°
1.5 kN

Question	Answer	Mark
	The magnitude of D is 1.92 kN.	**1**

Examiner's tip If you worked this out by scale drawing, some leeway would be allowed by the examiners; answers in the range 1.85 to 2.0 kN would be acceptable. The value of D can also be calculated using the sine rule; the angle opposite to D is 130° and that opposite to the 0.65 kN force is 15° so

$$\frac{D}{\sin 130^\circ} = \frac{0.65}{\sin 15^\circ}.$$

Question	Answer	Mark
(c) (i)	1.5 kN	**1**
	at 40° to the horizontal	**1**

Examiner's tip Ropes can only pull, not push; and they can only exert forces along their own direction.

Question	Answer	Mark
(ii)	1.5 kN × cos 40°= 1.15 kN	**1**
(iii)	The total opposing force = 2.35 kN	
	The work done each second = Fs = 2.35 kN × 8.5 m	**1**
	= 20.0 kJ	
	so the power is 20.0 kW	**1**
7 (a)	A vector has both size and direction, but a scalar only has a size.	**1**
(b) (i)	torque = force × perpendicular distance to pivot	**1**
	= 15 N × 15 cm (0.15 m)	
	= 225 Ncm (2.25 Nm)	**1**
(ii)	distance moved = $2\pi r = 2 \times p \times 0.15 = 0.94$ m	**1**
	work = force × distance	**1**
	= 15 N × 0.94 m = 14.1 J	**1**

Examiner's tip The distance must be measured in m for the answer to be in J. Although the 'rounded-off' answer for the distance was written down, it is a good idea to leave the unrounded value on your calculator display for use in subsequent calculations. This avoids rounding errors becoming magnified.

Question	Answer	Mark

(c) The excess pressure inside the bottle = force ÷ area **1**

$$= \frac{30\text{ N}}{\pi(1.0 \times 10^{-2})^2\text{ m}^2} = 9.5 \times 10^4\text{ Pa}$$

The total pressure needed is the sum of this and the atmospheric pressure that is pushing on the outside of the cork, i.e. 1.95×10^5 Pa **1**

> **Examiner's tip** This question shows the importance of using SI units in calculations. Length should be in m, mass in kg and time in s.

---

8 (a) The completed table is:

Quantity	Scalar	Vector
Displacement	☐	☑
Momentum	☐	☑
Energy	☑	☐
Magnetic flux density	☐	☑
Charge	☑	☐
Half-life	☑	☐

Award one mark for each two correct rows. **3**

(b) (i)

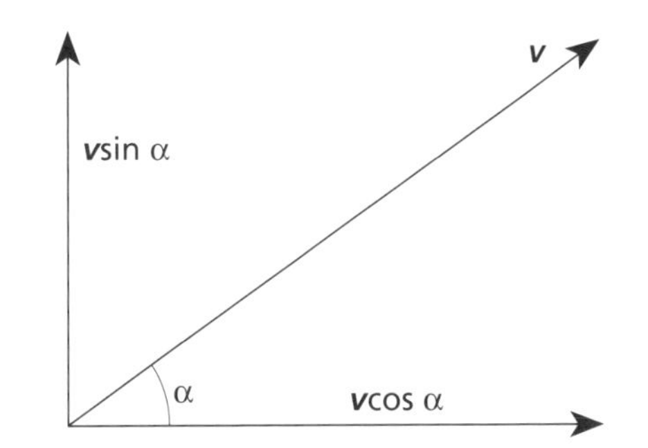

> **Examiner's tip** There is one mark for drawing the 'horizontal' and 'vertical' arrows the correct length, making the sides of a rectangle. The other two marks are for labelling the horizontal and vertical components correctly, 1 mark each. **3**

(ii) 90° **1**

(iii) 45° **1**

(c) (i)

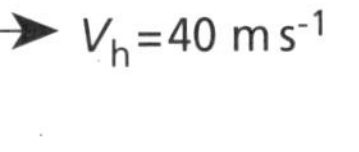

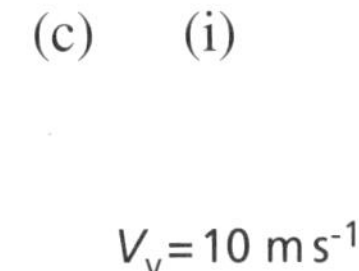

V = 41.2 m s^{-1}

3

> **Examiner's tip** One mark for each correct arrow and value.
> The stone maintains a constant velocity horizontally (ignoring air resistance), and accelerates vertically at 10 m s^{-2}, giving it a vertical velocity of 10 m s^{-1} after one second (from $v = u + at$ where $u = 0$ and $a = g$ and $t = 1$). The resultant velocity is found by adding these as vectors; as they are at right angles this is easily done using Pythagoras' theorem ($10^2 + 40^2 = V_R^2 = 1700$).

Question	Answer	Mark

Examiner's tip This part of the question uses the equations of uniform motion, which are covered in Unit 2.

Question	Answer	Mark
(ii)	For the vertical velocity,	
	initial velocity $u = 0$	
	acceleration $a = 10 \text{ m s}^{-2}$	
	distance travelled $s = 50 \text{ m}$	
	use $v^2 = u^2 + 2as$ to find v	1
	$v = \sqrt{2 \times 10 \times 50} = 31.6 \text{ m s}^{-1}$	1
(iii)	This diagram is for (iii) and (iv):	

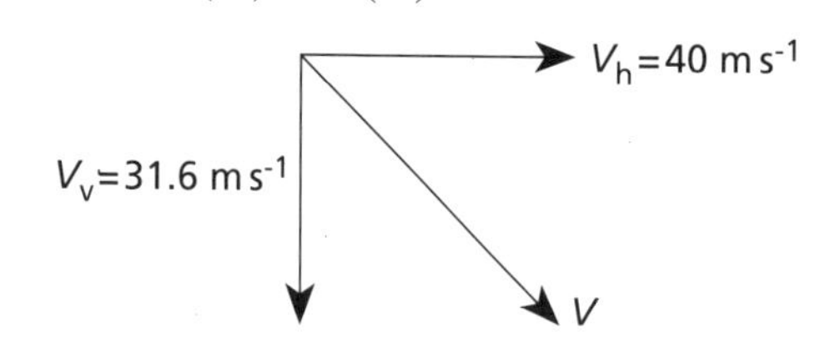

Examiner's tip The marks are awarded for:
'horizontal' line 40 units long — 1
'vertical' line 31.6 (or 32 to be realistic) units long — 1

Question	Answer	Mark
(iv)	51.0 m s^{-1}	1
	38° to the horizontal	1

2 FORCE AND MOTION

Question	Answer	Mark
1	C	1

Examiner's tip The equation to use is $v^2 = u^2 + 2as$.

Question	Answer	Mark
2	C	1

Examiner's tip The total force on the driver is the vector sum of the upwards force and the backwards force. The seat exerts an upwards force of 800 N to support the driver's weight. During braking, the backwards force, using $F = ma$, is equal to 200 N. Since these forces act at right angles, they are added using Pythagoras' theorem to give the total force on the driver.

Question	Answer	Mark
3	D	1

Examiner's tip The impulse, = force × time for which it acts, is represented by the area between the graph line and the time axis.

Question	Answer	Mark
4	**D**	**1**

Examiner's tip The resultant force required towards the centre is 10 N (using mv^2/r). The downward force (the weight of the ball) is 2 N so the tension needs to be 12 N.

Question	Answer	Mark
5 (a)	force × distance moved = work done	**1**
	force = 18 000 J ÷ 30 m	**1**
	= 600 N	**1**

Examiner's tip When the car is travelling at a constant speed, all the useful work is being done against resistive forces.

Question	Answer	Mark
(b)	driving force at 10 m s^{-1}= 18000 N ÷ 10 = 1800 N	**1**
	resistive force = 600 N ÷ 3 = 200 N	**1**
	resultant upwards force = 1600 N	**1**
	acceleration = $F \div m$ = 1600 N ÷ 1200 kg = 1.33 m s^{-2}	**1**

Question	Answer	Mark
6 (a) (i)	momentum = mass × velocity	**1**
	= 0.06 kg × 48 m s^{-1} = 2.88 N s	**1**

Examiner's tip The mass needs to be in kg and the velocity in m s^{-1} for the answer to be in N s. An alternative unit for momentum is kg m s^{-1}.

Question	Answer	Mark
(ii)	force = momentum change ÷ time	**1**
	= 2.88 ÷ (8×10^{-3}) = 360.0 N	**1**
(iii)	acceleration = increase in velocity ÷ time	
	= 48 ÷ (8×10^{-3}) = 6×10^3 m s^{-2}	**1**
(b)	The completed graph is shown below.	

The curve is steep at 4 ms and gets less steep towards 8 ms. **1**
The gradient decreases to zero after 8 ms. **1**
The speed is maintained at 48 m s^{-1} for the last 2 ms. **1**

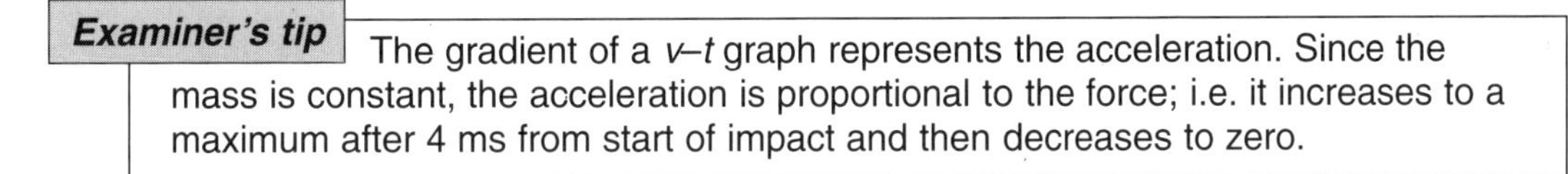
Examiner's tip The gradient of a v–t graph represents the acceleration. Since the mass is constant, the acceleration is proportional to the force; i.e. it increases to a maximum after 4 ms from start of impact and then decreases to zero.

Question			Answer	Mark
	(c)		The distance travelled is represented by the area between the graph line and the time axis.	1
			The number of squares can be estimated; each small square on the graph represents a distance of 2 m s^{-1} × 0.4 ms = 0.8 mm.	1
7	(a)		Inelastic.	1

Examiner's tip A collision is inelastic whenever work is done against internal forces; in this case the pellet does work in deforming the car when it becomes embedded.

Question			Answer	Mark
	(b)		The initial speed of the car is 2 m s^{-1}.	1

Examiner's tip This follows from conservation of momentum; all the momentum of the pellet has been transferred to a combined mass 100 times as great as the mass of the pellet, so the velocity is 1/100 as the momentum remains the same.

Question			Answer	Mark
			As the car rises up the slope, ke → gpe, i.e. $^1/_2 mv^2 = mg\Delta h$	1
			$\Delta h = {^1/_2} v^2/g$	1
			$= {^1/_2} \times 2^2/10 = 0.2$ m	1

Examiner's tip A common error in questions of this type is to assume that the kinetic energy of the pellet becomes gravitational potential energy of the car. This would lead to an answer of 20 m, which shows that a large proportion of the kinetic energy of the pellet becomes internal (thermal) energy of the car.

Question			Answer	Mark
8	(a)		The weight of the spacecraft = 15 000 kg × 1.6 N kg^{-1} = 24 000 N	1
			The resultant force is zero, so there is no change in velocity.	1
	(b)	(i)	resultant upward force = 1500 N	1
			deceleration = $F \div m$	1
			= 1500 ÷ 15 000 = 0.1 m s^{-2}	1
		(ii)	change in speed during deceleration = 0.1 m s^{-2} × 18 s = 1.8 m s^{-1}	1
			landing speed = 2 – 1.8 = 0.2 m s^{-1}	1
		(iii)	average speed during deceleration = $^1/_2$(2 + 0.2) = 1.1 m s^{-1}	1
			distance travelled in this time = 1.1 m s^{-1} × 18 s = 19.8 m	1

Examiner's tip The last part of the question could be answered using $v^2 = u^2 + 2as$ or $s = ut + {^1/_2}at^2$.

Question			Answer	Mark
9	(a)		Angular velocity is the angle turned through per second or the rate of change of angular displacement.	2
	(b)	(i)	$v = r\omega$	1
		(ii)	v can be varied by changing the length of the cord.	1
		(iii)	The stone is accelerating towards the centre of the circle.	1
			An unbalanced force is needed towards the centre of the circle to cause this acceleration.	1

Question			Answer	Mark
			The string tension provides this force by pulling on the stone.	1
	(iv)		acceleration, $a = v^2/r$	1
			therefore the force T (using $F = ma$) $= mv^2/r$	1
			replace one v in the v^2 by $r\omega$ from (b)(i)	
			gives $T = mvr\omega/r = mv\omega$	1
(c)	(i)	(1)	the centripetal acceleration $= v^2/r = 12^2/7 = 20.6$ m s^{-2}	1
		(2)	using $F = ma$, the resultant force needed to cause this acceleration is $60 \times 20.6 = 1234$ N (or 1236 N if you used the unrounded value of $12^2/7$)	1
			The weight of the passenger = 60 kg × 10 N kg^{-1} = 600 N	1
			The seat must exert a force of 1234 – 600 = 634 N (or 1236 – 600 = 636 N)	1

Examiner's tip Remember, the centripetal force is the sum or resultant of the forces that are acting. In this case both forces, the weight of the passenger and the force from the seat, act downwards, towards the centre of the circle.

Question		Answer	Mark
(ii)	(1)	change in gpe $= mg\Delta h$	1
		$= 60 \times 10 \times 14 = 8400$ J	1
	(2)	assuming all the gpe is transferred to ke,	
		ke gain $= \frac{1}{2}mv^2 = 8400$ J; initial ke = 4320 J	1
		∴ total ke = 12720 J	1
		speed at bottom $= \sqrt{2 \times 12720 \div 60} = 20.6$ m s^{-1}	1

Examiner's tip When objects change their position above the surface of the Earth, the energy transfer is from ke to gpe for an ascending object and from gpe to ke for a descending object.

Question	Answer	Mark
(iii)	The speed must be high enough so that, after the passenger has slowed down going up the loop, the centripetal acceleration is at least equal to g, otherwise the passenger will fall out of the ride.	1

Examiner's tip If the centripetal acceleration at the top of the loop is less than g, the passenger's downwards acceleration is greater than that required for circular motion; hence the passenger accelerates downwards at a greater rate than the car does — they part company.

3 VIBRATIONS AND WAVES

Question	Answer	Mark
1	**B**	**1**

Examiner's tip A and B are examples of things vibrating at their natural frequency without a forcing vibration. D is a forced vibration but not necessarily at the natural frequency.

Question	Answer	Mark
2	B	1

Examiner's tip The highest frequency wave has the shortest wavelength.

3	C	1

Examiner's tip This is calculated using Snell's law,

$$\frac{\sin i}{\sin r} = \frac{v_1}{v_2}$$

where i = angle of incidence, r = angle of refraction

4	E	1

Examiner's tip To arrive at this answer, use the diffraction grating formula $n\lambda = d \sin \theta$. The maximum value of $\sin \theta$ is 1, so

$$n \leq \frac{d}{\lambda}$$

This gives a maximum value for n of 3.2. Since n can only have integer values, three transmitted beams (for non-zero angles θ) can occur in each quadrant. Hence, together with the straight through beam ($n = 0$), we have 7 beams.

---

5 (a) From the diagram, $x = 10.8 \div 12 = 0.9$ cm (9×10^{-3} m). **1**

Rearrange the equation to give $d = \frac{\lambda D}{x}$ **1**

$= (6.33 \times 10^{-7} \times 3.02) \div 9 \times 10^{-3}$

$= 2.12 \times 10^{-4}$ m **1**

Examiner's tip Notice how all the distances have been changed into metres to avoid confusion. Notice also potential confusion between the meanings of d and D.

(b) The dots will become closer together. **1**

Examiner's tip This happens because the wavelength of green light is less than that of red light, and the spacing of the maxima (the dots in this case) is proportional to the wavelength of the light used.

---

6 (a)

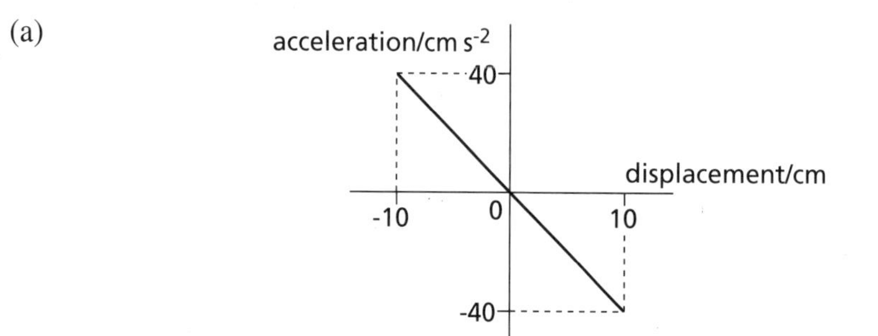

Question	Answer	Mark

Examiner's tip Many candidates make the error of drawing a sine curve when they see the words 'simple harmonic motion'. Acceleration in simple harmonic motion varies with displacement according to the equation $a = -\omega^2 x$, where ω is a positive constant; so acceleration is proportional to displacement, but always in the opposite direction, hence the line has a negative slope.

Answer	Mark
The four marks are awarded for:	
straight line through the origin	**1**
negative gradient	**1**
maximum displacement labelled as '10 cm'	**1**
maximum acceleration labelled as '40 cm s^{-2}'	**1**

Examiner's tip The maximum displacement is the amplitude given in the question. The maximum acceleration is calculated using the equation $a = -\omega^2 x$. Since $\omega = 2\pi f$, it follows that $\omega = 2$ and $\omega^2 = 4$, so the maximum acceleration is numerically equal to four times the maximum displacement. As x is in cm it follows the acceleration will be in cm s^{-2}.

(b)

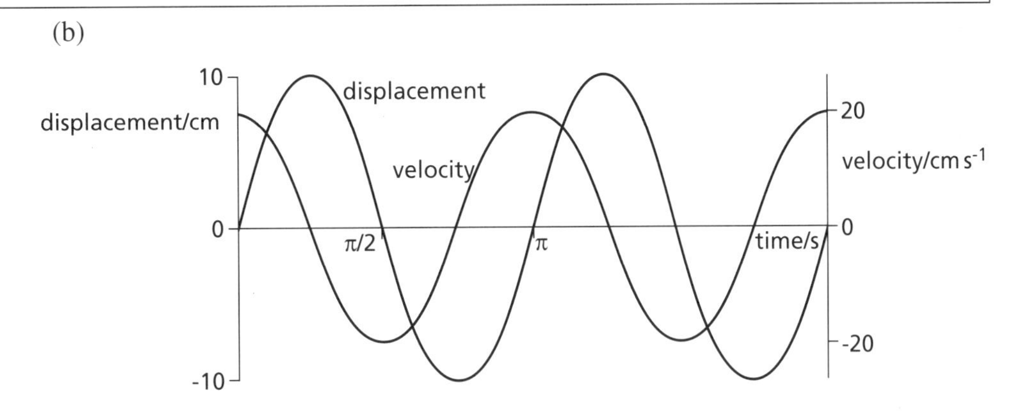

Examiner's tip The graphs are drawn according to the solutions to the shm equation, $x = a \sin \omega t$ and $v = a\,\omega \cos \omega t$. Note the 2 vertical scales: one for displacement, the other for velocity.

Answer	Mark
The marks are awarded for:	
Two sinusoidal curves	**1**
Correct phase relationship	**1**
Correct numerical values on time axis	**1**
Correct numerical values on displacement and velocity axes	**1**

Examiner's tip The period $T = 1/F$ where $F = 1/\pi$. The maximum speed $v = a\omega$, $= 10 \times 2$, $= 20$ cm s^{-1}.

Question			Answer	Mark
7	(a)	(i)	The angle of refraction is 30°	**1**
			$\mu = \dfrac{\sin i}{\sin r} = \dfrac{\sin 40°}{\sin 30°} = 1.29$	**1**
		(ii)	The ray would have a smaller angle of refraction at Q, i.e. it would be deviated through a larger angle.	**1**
	(b)	(i)	$\sin c = \dfrac{1}{\mu} = \dfrac{1}{1.80} = 0.556$	**1**
			$c = \sin^{-1} 0.556 = 33.7°$	**1**

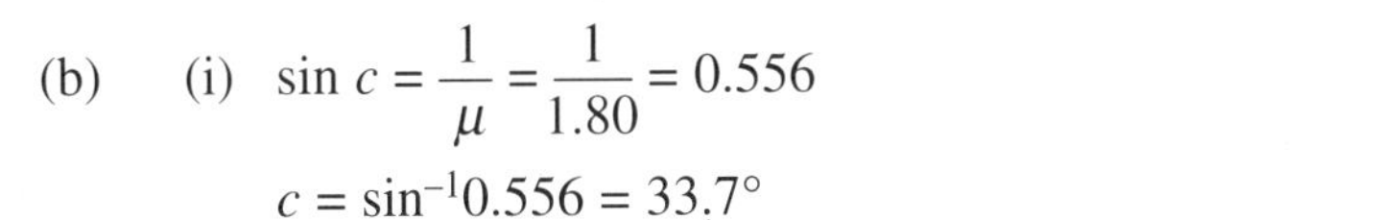

Question	Answer	Mark
(ii)	The angle of refraction at the first face is 21°.	1
	The angle of incidence at the second face is 39°, so the ray is reflected internally at the same angle (as this exceeds the critical angle).	1
	The ray is next incident on the bottom face at an angle of incidence of 21° so it emerges at an angle of refraction of 40°.	1

Examiner's tip For the last refraction sin 21°/sin r = 1/1.8, as the light is travelling from glass to air.

Question	Answer	Mark
8 (a) (i)	The variation is due to interference between the waves from both sources.	1
	At a place where the waves arrive in phase there is a maximum.	1
	The minima occur at places where the waves arrive with a path difference of half a wavelength.	1

Examiner's tip When explaining interference phenomena you should always try to identify the places where constructive and destructive interference occurs.

Question	Answer	Mark
(ii)	The wavelength, $\lambda = v/f = 340 \div 3400 = 0.1$ m	1
	The separation of the maxima $x = \frac{D\lambda}{d} = \frac{9.0 \times 0.1}{1.5} = 0.6$ m	
	So the distance to the first minimum is half this, i.e. 0.3 m	1
(b)	The wavelength would decrease.	1
	This would cause the distance to the first minimum to decrease.	1

Examiner's tip Note that the frequency of the sound source would not be affected, so the wavelength is proportional to the speed of the waves.

Question	Answer	Mark
9 (a) (i)	A ultra-violet	1
	C infra-red	1
(ii)	The photoelectric equation is $hf = hc/\lambda = W + (^1/_2mv^2)_{max}$	1
	For metal P, the threshold wavelength, i.e. the wavelength for which $hc/\lambda = W$, lies between 600 and 800 nm.	1
	For metal Q, the threshold wavelength lies between 300 and 600 nm.	1
	The emission of electrons only occurs at wavelengths shorter than the threshold.	1

Examiner's tip Photoelectric emission only occurs when the energy of each photon is at least equal to the minimum amount of energy needed to liberate an electron. This minimum amount of energy is called the work function (W) and is different for different metals. Since energy is proportional to frequency and inversely proportional to wavelength, there is a cut-off at a minimum frequency and a maximum wavelength which will cause photoelectric emission.

Question	Answer	Mark
(iii)	The results would not change	1
	since doubling the intensity does not affect the wavelength or frequency	1
	of the light, it only changes the rate at which photons are emitted.	1
(b) (i)	eV is the energy transfer from the electric field to the electron	
	$^1/_2mv^2$ is the kinetic energy gained by the electron	1

Question	Answer	Mark
	so the equation states that the energy lost by the field is equal to the energy gained by the electron.	1
(ii)	$v = \sqrt{\dfrac{2eV}{m}} = \sqrt{\dfrac{2 \times 1.60 \times 10^{-19} \times 2000}{9.11 \times 10^{-31}}}$	1
	$= 2.65 \times 10^{7}\ \mathrm{m\,s^{-1}}$	1
(c) (i)	momentum of the electrons, $p = mv = 9.11 \times 10^{-31} \times 2.65 \times 10^{7}$ $= 2.41 \times 10^{-23}$ Ns	1
	$\lambda = h/p = 6.6 \times 10^{-34} \div 2.41 \times 10^{-23} = 2.7 \times 10^{-11}$ m	1
(ii)	The spaces between the atoms in graphite are of the same order of magnitude as the wavelength calculated in (i).	1

Examiner's tip You are not expected to have detailed knowledge of the spacing between the atoms in graphite, but you are expected to apply your knowledge of the conditions under which waves are diffracted through large angles to this situation.

Question	Answer	Mark
(iii)	Apply a magnetic field (or electric field)	1
	at right angles to the path of the particles.	1

magnetic field acting into paper

OR electric field acting towards the top of the paper

Question	Answer	Mark
	The diagram shows the deflection of the particles if the particles are negatively charged.	1
	This would cause the rings to be formed lower on the screen.	1

Examiner's tip Your test should be based on deflecting the particles by using either an electric or a magnetic field. It is easier and safer to use a magnetic field, since this can be done with a permanent magnet whereas an electric field would require a large potential difference that could be a safety hazard.

---

10 (a) (i)

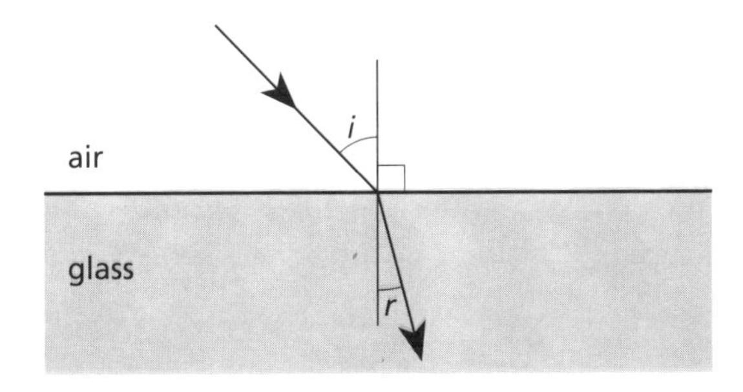

Question	Answer	Mark
	The marks are awarded for:	
	Correct change in direction (towards the normal in glass)	1
	Angle i marked correctly	1
	Angle r marked correctly	1

Question	Answer	Mark
(ii)	Total internal reflection can occur at a boundary where the waves would speed up, e.g. light passing from glass into air.	1
	Above a certain angle of incidence (the critical angle), all of the light is reflected.	1
(iii)	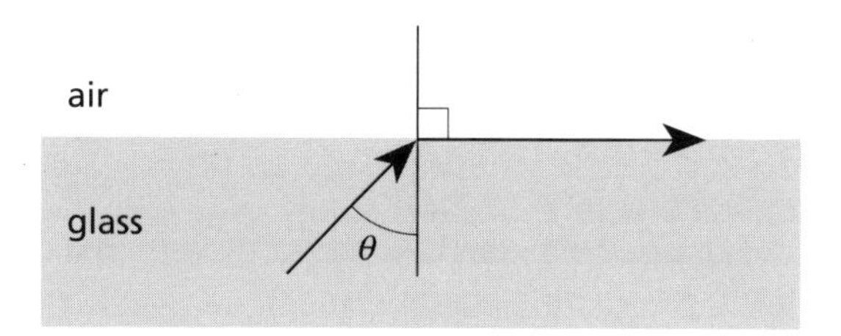	
	A diagram showing the critical angle being when the light leaves the glass at an angle of 90°	1
	$n = \frac{\sin 90°}{\sin \theta}$ OR $\frac{1}{n} = \frac{\sin \theta}{\sin 90°}$	1
	as $\sin 90° = 1$, $\sin \theta = \frac{1}{n}$,	1

Examiner's tip

Take care to avoid confusion over the relationship

$$n = \frac{\sin i}{\sin r}.$$

When written in this form, it applies to light being slowed down, e.g. when passing from air or a vacuum into another substance. The ratio of the sines of the angles for light following the reverse path is $\frac{1}{n}$. The first equation treats the light as if it were passing from air to glass with an angle of incidence of 90° and an angle of refraction of θ. The second equation uses θ as the angle of incidence and 90° as the angle of refraction, with the ratio of their sines being $\frac{1}{n}$.

Question	Answer	Mark
(b) (i)	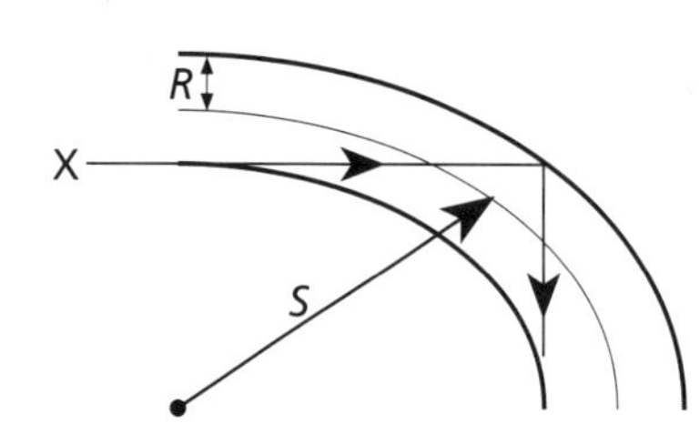	
	Your diagram should show X being reflected internally	1
	With the angles of incidence and reflection being equal	1
(ii)	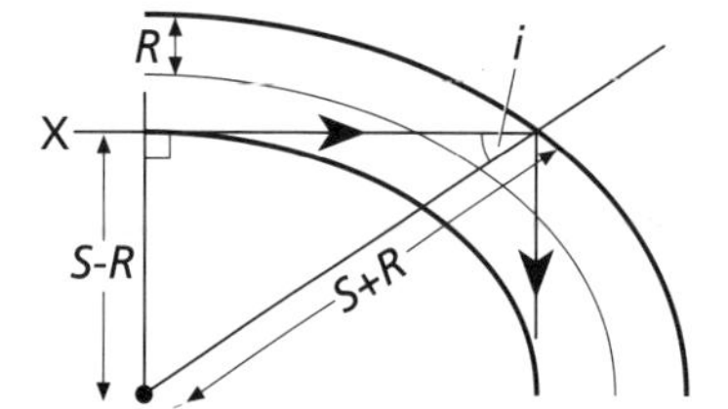	

Examiner's tip

The diagram shows the angle of incidence with two radial lines drawn in. A radius of a circle meets the circumference at a right angle.

Question	Answer	Mark
	Construction of right-angled triangle	1
	The side opposite to i has length $S–R$ and the hypotenuse of the triangle has length $S+R$	1
	Hence $\sin i = \frac{\text{opposite}}{\text{hypotenuse}} = \frac{S-R}{S+R}$	1
	Light can escape if this angle is less than the critical angle i.e. if $\frac{S-R}{S+R} < \frac{1}{n}$	1
	inverting this equation gives $n < \frac{S+R}{S-R}$	1

Examiner's tip When an equation with an inequality is inverted, the inequality is reversed.

Question	Answer	Mark
(c) (i)	Time for radio waves to travel 50 km $= 5 \times 10^4$ m $\div 3 \times 10^8$ m s^{-1} $= 1.67 \times 10^{-4}$ s	1
	The speed of light in glass $= \frac{3 \times 10^8}{1.5} = 2 \times 10^8$ m s^{-1}	1
	Time for light to travel in glass $= 5 \times 10^4 \times 2 \times 10^8 = 2.5 \times 10^{-4}$ s so the time difference is 0.83×10^{-4} s	1
(ii)	Glass fibre is preferred where there are hills (or where the distance is greater than 50 km).	1
	The microwaves would be blocked by hills/ would not follow the Earth's curvature.	1

Examiner's tip There are other possible answers to this question. The higher frequency of light enables data to be transmitted more rapidly in glass fibres than when using microwaves. Also, changes in atmospheric conditions would not affect glass fibres but could affect microwaves.

4 GRAVITATIONAL AND ELECTRICAL FIELDS

Question	Answer	Mark
1	D	1

Examiner's tip The force is inversely proportional to the (distance)^2, or $F \propto \frac{1}{d^2}$, so a graph of F against $1/d^2$ is a straight line through the origin.

Question	Answer	Mark
2	B	1

Examiner's tip Gravitational potential is taken to be zero at infinity. A satellite in orbit has less energy than an object out of the influence of the Earth's gravitational field, so it has a negative amount of energy. In moving to a higher orbit the satellite gains gravitational potential energy, but it still has a negative value.

Question	Answer	Mark
3	E	1

Examiner's tip When the charged particle is stationary the forces from the electric field and the Earth's gravitational field are balanced. For the particle to move downwards, either the electric force (pulling it upwards) must decrease or the gravitational force must increase.

Question	Answer	Mark
4	D	1

Examiner's tip The field strength is equal to –(potential gradient). The gradient of the potential-distance graph is positive and constant so the field strength is negative and constant.

---

5 (a) The gravitational force is the unbalanced force required for circular motion,

i.e. $\frac{mv^2}{r} = \frac{GMm}{r^2}$ **1**

$v^2 = \frac{GMmr}{mr^2}$ so $v = \sqrt{\frac{GM}{r}}$ **1**

(b) From (a), $M = \frac{v^2 r}{G}$ **1**

Examiner's tip To derive this, first square each side of the equation; next multiply each side by *r* and divide each side by *G*.

Substitute the values for an outermost particle:

$M = \frac{(1.7 \times 10^4)^2 \times 1.4 \times 10^8}{6.7 \times 10^{-11}} = 6.0 \times 10^{26}$ kg **1**

(c) time = distance ÷ speed $= \frac{2\pi \times 1.4 \times 10^8}{1.7 \times 10^7}$ **1**

$= 5.17 \times 10^4$ s **1**

(d) From (a), $v = \sqrt{\frac{GM}{r}} = \sqrt{\frac{6.7 \times 10^{-11} \times 6.0 \times 10^{26}}{7 \times 10^7}} = 2.40 \times 10^4$ m s^{-1} **1**

---

6 (a) (i) $F = mg$ **1**

$F = \frac{GMm}{r^2}$ **1**

(ii) Rearranging and combining the equations gives

$M = \frac{Fr^2}{Gm} = \frac{mgr^2}{Gm} = \frac{gr^2}{G}$ **1**

(b) density = mass ÷ volume **1**

$= \frac{M}{V} = \frac{3gr^2}{4\pi r^3 G} = \frac{3g}{4\pi rG}$ **1**

Question	Answer	Mark
	$= \dfrac{3 \times 9.81}{4 \times \pi \times 6.4 \times 10^{6} \times 6.7 \times 10^{-11}} = 5.46 \times 10^{3}$ kg m^{-3}	1

Examiner's tip It is simpler to manipulate the symbols first to give an expression for density before substituting in values.

(c)	The Earth's poles are 'flat'.	1
	So the distance to the centre of the Earth is less, making g greater.	1

7	The diagram is shown below:	

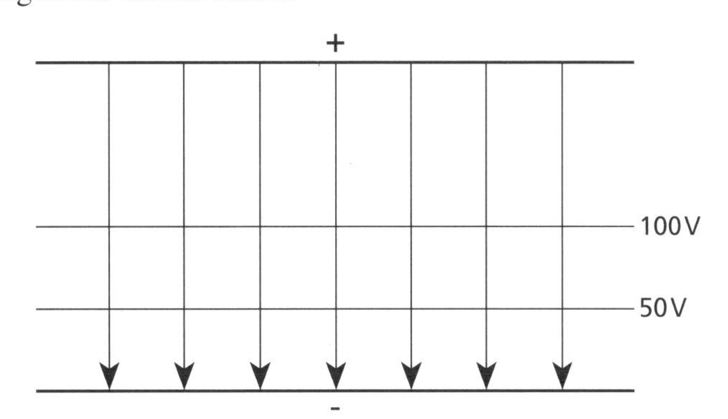

	Marks are awarded for:	
	Field shown as straight lines between plates	1
	+ and – plates marked	1
	arrows go from + to –	1
	100 V equipotential parallel to the plates and halfway between them	1
	50 V equipotential 1/4 of the plate separation from the – plate	1

Examiner's tip Equipotentials are lines joining points at the same potential. Since the electric field between the plates has a uniform value, it follows that the potential changes at a uniform rate between the plates.

---

8 (a) (i)	The electric potential at a point is the work done per unit positive charge in bringing a small charge from infinity to that point.	2

Examiner's tip In this definition, the test charge has to be small so that its field does not affect the field under test. As with gravitational fields, 'infinity' means far enough away for the forces due to the field to be negligible.

(ii)	$E = \dfrac{Qq}{4\pi\varepsilon_0 r}$	1

Examiner's tip Since the potential is the energy per unit charge, the energy is calculated simply by multiplying the potential by the amount of charge.

(b)	The closest distance of approach occurs when the kinetic energy of the alpha particle has been transferred to potential energy of the field.	1
	i.e. $\frac{1}{2}mv^2 = \dfrac{Qq}{4\pi\varepsilon_0 r}$	1

Question	Answer	Mark
	$r = \dfrac{2Qq}{4\pi\varepsilon_0 mv^2}$	1
	$= \dfrac{2 \times 3.2 \times 10^{-19} \times 9.0 \times 10^9}{6.8 \times 10^{-27} \times (1.2 \times 10^7)^2}$	1
	$= 6.59 \times 10^{-15}$ m	1

9 (a)

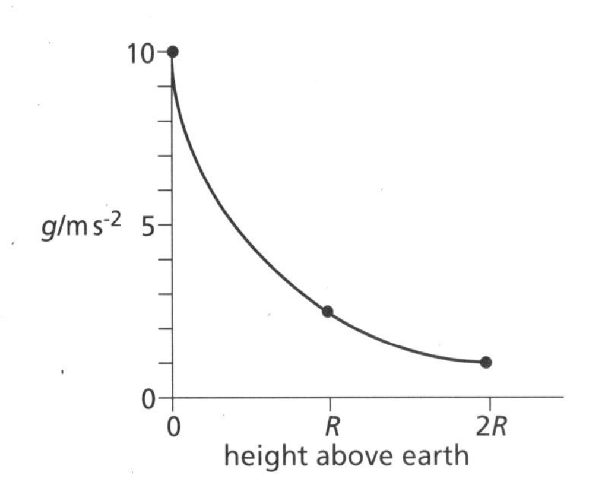

The marks are awarded for:

	Axes marked with values and labels	1
	Correct curved shape of line	1
	g = approx. 10 at $h = 0$	1
	$g = 2.5$ at $h = R$ and 1.1 at $h = 2R$	1

Examiner's tip The value of g is inversely proportional to the square of the distance from the centre of the Earth. At the Earth's surface this distance is R and at a height R above the surface the distance from the Earth's centre is $2R$. In moving from the surface to a height of R this distance has been doubled, so g has reduced to one quarter of its value at the Earth's surface. Similarly, at a height of $2R$, g is one ninth of its value at the Earth's surface since it is three times as far from the Earth's centre.

(b) (i) ke = $\frac{1}{2}mv^2$

$57.8 \times 10^{10} = \frac{1}{2} \times 40\,000 \times v^2$ — **1**

$v = \sqrt{\dfrac{2 \times 57.8 \times 10^{10}}{40\,000}} = 5380 \text{ m s}^{-1}$ — **1**

(ii) work done = energy transfer from kinetic energy

$= 5.7 \times 10^{10}$ J — **1**

Examiner's tip The expression 'work done' means the same as 'energy transfer'. In this case the energy transfer is from kinetic energy to gravitational potential energy.

10 (a) The gravitational field strength is the force acting per unit mass. — **1**

(b) gravitational field strength = 9.8 N kg^{-1} (1 mark for the value and 1 for the unit) — **2**

(c) g is the value of free-fall acceleration. — **1**

Question		Answer	Mark
		The acceleration is caused by gravitational forces; it is not a gravitational force.	1
(d)	(i)	speed = distance ÷ time	1
		$= 2\pi r \div t = \dfrac{2\pi \times 3.84 \times 10^8}{2.36 \times 10^6} = 1.02 \times 10^3$ m s^{-1}	1
	(ii)	acceleration = v^2/r	1
		$= \dfrac{(1.02 \times 10^3)^2}{3.84 \times 10^8} = 2.72 \times 10^{-3}$ m s^{-2}	1
	(iii)	force = mass × acceleration = $7.35 \times 10^{22} \times 2.72 \times 10^{-3} = 2.00 \times 10^{20}$ N	1
	(iv)	gravitational field strength = force ÷ mass = 2.72×10^{-3} N kg^{-1}	1

Examiner's tip This question shows that gravitational field strength and free fall acceleration have the same numerical values. Although they have different definitions, they are equivalent expressions since
force = mass × acceleration and
(gravitational) force = mass × gravitational field strength.

5 CURRENTS IN CIRCUITS

Question	Answer	Mark
1	E	1

Examiner's tip The cell's internal resistance acts as a series resistor in the circuit. The potential difference across the 20 Ω resistor is 1.0 V, so that across the cell resistance is 0.5 V. Since the same current passes through the two resistors, it follows that the internal resistance has half the value of the external resistance.

Question	Answer	Mark
2	C	1

Examiner's tip The power supplied by the battery is equal to I^2R. The total circuit resistance is 12 Ω so $I^2 = 0.01$, i.e. $I = 0.1$ A.

Question	Answer	Mark
3	D	1

Examiner's tip The effective capacitance between P and S is 8 μF plus the capacitance of the series combination of the other three capacitors. When capacitors are placed in series, the effective capacitance is always smaller than the smallest capacitor in the combination.

Question	Answer	Mark
4	B	1

Examiner's tip The charge and energy are calculated using $Q = CV$ and $E = \frac{1}{2}CV^2$.

Question	Answer	Mark
5	Connected in parallel means that each lamp is connected directly to the battery (and therefore each has 12 V across it).	1
	The current in each lamp does not pass through any other lamp.	1
	This enables each lamp to be used independently of the others.	1
	$I = P \div V$	1
	$= 144\text{ W} \div 12\text{ V}$ (from total power $= (2 \times 60) + (4 \times 6) = 144$ W)	1
	$= 12$ A	1

Examiner's tip The total power (144 W) is at a p.d. of 12 V hence the total current is 12 A. Alternatively there is 5 A in each of the 60 W bulbs and 0.5 A in each of the 6 W bulbs. Hence the total current is 12 A.

Question	Answer	Mark
6 (a)	Electromotive force is the total energy transfer per unit charge from the power supply.	1

Examiner's tip Do not be confused by the term electromotive force; it does not describe a force at all, but an energy transfer per unit charge.

Question	Answer	Mark
(b) (i)	Using $P = I^2R$, $I = \sqrt{\dfrac{P}{R}}$	1
	$= \sqrt{\dfrac{8}{0.32}} = 5.0$ A	1

Examiner's tip A common error here is to use the power supply e.m.f. to calculate the current using $I = P \div V = 4$ A. Because of the internal resistance of the power supply, the p.d. across the heater is less than 2 V.

Question	Answer	Mark
(ii)	$V = IR = 5.0\text{ A} \times 0.32\ \Omega = 1.6$ V	1
(iii)	The potential difference across the internal resistance $= 2.0\text{ V} - 1.6\text{ V} = 0.4$ V	1
	its resistance, $R = \dfrac{V}{I} = \dfrac{0.4}{5} = 0.08\ \Omega$	1

Examiner's tip Note that the current in the cell is the same as the current in the rest of the circuit.

Question	Answer	Mark
(c)	It will increase but less than double, because the potential difference across the cell terminals drops as more current is drawn from it.	1
	The total circuit resistance is now 0.24 Ω.	1

Examiner's tip Two heaters in parallel have half the resistance of one heater; i.e. 0.16 Ω. The resistance of the cell is added to this to find the total resistance of the circuit.

The current in the circuit is now $\dfrac{V}{R} = \dfrac{2.0}{0.24} = 8.33$ A so the total power in the heaters is equal to $I^2R = 8.33^2 \times 0.16 = 11.1$ W — **1**

Question	Answer	Mark
7 (a) (i)	By using an area of A and a separation of $2d$	1

Question	Answer	Mark

Examiner's tip The capacitance of a parallel-plate capacitor is given by the expression $C = \frac{\varepsilon A}{d}$; this has smallest value for the smaller value of A and the larger value of d.

(ii) It increases the capacitance. **1**

(iii) (A) $C = \frac{Q}{V}$ **1**

$= \frac{6 \times 10^{-9}\ \text{C}}{20\ \text{V}} = 3 \times 10^{-10}\ \text{F}$ **1**

Examiner's tip The same result would be obtained using the value of 3×10^{-9} C for the charge when the p.d. was 10 V.

(B) $E = \frac{1}{2}QV$ (or $\frac{1}{2}CV^2$) **1**

$= \frac{3 \times 10^{-9} \times 10}{2} = \left(\text{OR} = \frac{3 \times 10^{-10} \times 10^2}{2}\right) = 1.5 \times 10^{-8}\ \text{J}$ **1**

(C) $E = \frac{6 \times 10^{-9} \times 20}{2} = 6.0 \times 10^{-8}\ \text{J}$ **1**

(b) (i) Above a certain voltage the insulator between the plates breaks down and conducts electric current. **1**

(ii) Rearranging $E = \frac{V}{d}$ gives $d = \frac{V}{E}$ **1**

$= \frac{400}{30\ 000} = 1.33 \times 10^{-2}\ \text{cm}$ **1**

Examiner's tip Note that, as the E field is in kV cm^{-1} the separation is in cm on substitution.

(c) (i) $I = \frac{V}{R} = \frac{8\ \text{V}}{22 \times 10^3\ \Omega} = 3.64 \times 10^{-5}\text{A}$ **1**

Examiner's tip Immediately the switch is closed, there is no p.d. across the capacitor so the current is determined by the supply p.d. and the value of the resistance in the circuit.

(ii) (A) Shape of graph **1**
maximum p.d. approaches 8 V **1**

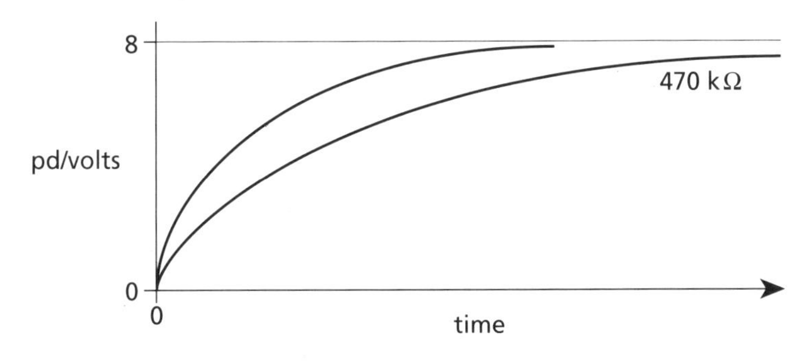

Question	Answer	Mark

Examiner's tip The graph shows the p.d. rising rapidly at first, with the rate of increase of p.d. then decreasing. This occurs because the charging current decreases as the p.d. across the capacitor rises, causing the resultant p.d. across the resistor to fall.

(B) The graph shows the p.d. increasing at a reduced rate. **1**

Examiner's tip The effect of increasing the resistor is to reduce the charging current, so the capacitor takes longer to charge to any given value of p.d.

(iii) It increases. **1**

Examiner's tip When an alternating p.d. is applied to a capacitor, the capacitor is repeatedly charged, discharged and then charged with the opposite polarity. Increasing the frequency increases the number of times this happens each second, so increasing the rate of flow of charge, i.e. the current.

---

8 (a) (i) Negatively charged particles (electrons) move onto the plate connected to the negative side of the power supply, giving this plate a negative charge. **1**
Electrons move away from the plate connected to the positive side of the power supply, giving this plate a positive charge. **1**
The graph is a straight line through the origin (see below). **2**

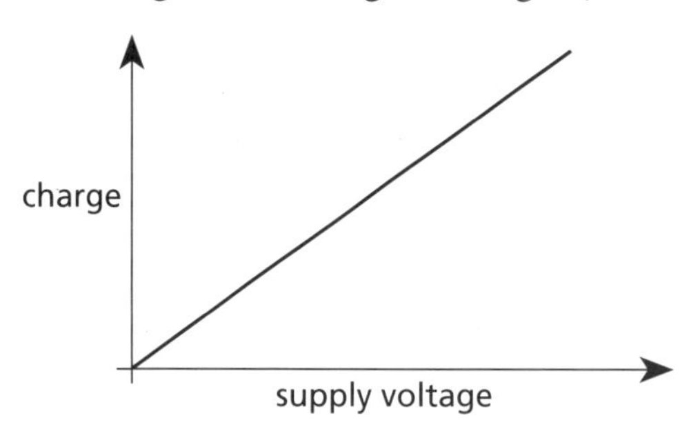

To obtain the value for the capacitance:
draw in the line of best fit **1**
the gradient of this line represents the capacitance **1**

Examiner's tip A calculation based on the gradient is preferred to working out the capacitance from a pair of values, as this gets rid of the effect of any zeroing errors that would affect experimental values but would not affect the gradient of the line.

$E = \frac{1}{2}CV^2$ **1**
Capacitors store small amounts of energy relative to their physical size **1**
so the space available is a limiting factor. **1**
There is a limit to the voltage that can be applied before the insulator breaks down and conducts electricity. **1**

(ii)

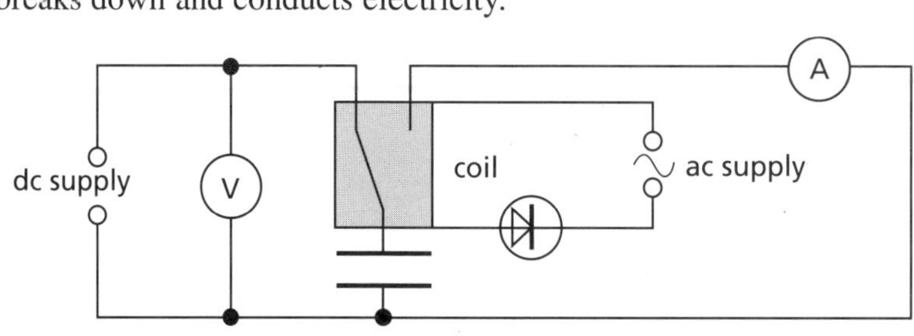

Question	Answer	Mark
	Marks are awarded for:	
	a.c. supply to coil (surrounding magnetic reed switch)	1
	charging circuit (left hand part of circuit shown on diagram)	1
	discharging circuit correct (right hand part of circuit shown on diagram)	1

Examiner's tip Note that the diode is not absolutely necessary. Using the diode makes the reed switch vibrate at the same frequency as the a.c. supply used; this could be from a signal generator or a low voltage derived from the mains using a transformer. If the diode is not used the reed vibrates at twice the frequency of the a.c. supply.

To calculate the capacitance:
record the d.c. supply voltage, V — 1
and the a.c. supply frequency, f — 1
and the discharge current, I — 1

The capacitance, $C = \frac{Q}{V} = \frac{I}{fV}$ — 1

Examiner's tip The current, *I*, is the charge flow per second, i.e. the charge flow due to *f* discharges of the capacitor. Therefore the charge stored on the capacitor plates is *I/f*. If a diode is not used in the circuit this becomes *I/2f*.

(b) (i) $E = \frac{1}{2}CV^2$ — 1

$= \frac{1}{2} \times 20 \times 10^{-6} \times 12^2 = 1.44 \times 10^{-3}$ J — 1

(ii) The charge initially on the 20 μF capacitor,
$Q = CV = 20 \times 10^{-6} \times 12 = 2.4 \times 10^{-4}$ C — 1
This is shared between the capacitors. As they end up at the same p.d., the 20 μF capacitor will have twice the charge of the 10 μF capacitor. — 1
It will therefore have two thirds of the charge, i.e. 1.6×10^{-4} C — 1

so its p.d., $V = \frac{Q}{C} = \frac{1.6 \times 10^{-4}}{20 \times 10^{-6}} = 8$ V — 1

Examiner's tip Award yourself full marks if you arrived at the correct answer without the numerical calculations, which are not strictly necessary. It is correct to work only in terms of ratios with this question; since the charge on the 20 μF capacitor is reduced to two thirds of its initial value, it follows that the p.d. changes in the same ratio.

(iii) Final energy stored $= \frac{1}{2}QV = \frac{2.4 \times 10^{-4} \times 8}{2} = 9.6 \times 10^{-4}$ J — 1

Examiner's tip An alternative way to calculate this is to use $\frac{1}{2}CV^2$ with *C* for the two capacitors in parallel equal to 30 μF.

The difference in the energies is 4.8×10^{-4} J. — 1
Connecting the capacitors together caused charge to oscillate between the capacitors. — 1
This oscillation was damped by the resistance of connecting wires — 1
causing the energy to be transferred as heat in the wires. — 1

Question	Answer	Mark

Examiner's tip The oscillation is caused by there being more energy than that associated with the new equilibrium state.

6 ELECTROMAGNETISM

Question	Answer	Mark
1	B	1

Examiner's tip Conductors that carry currents in the same direction attract each other and those carrying currents in opposite directions repel each other.
Both magnetic forces on Y are directed towards X.

Question	Answer	Mark
2	B	1

Examiner's tip The magnetic field pattern due to a current-carrying conductor is in the form of concentric circles around the conductor. These point in a clockwise direction when viewed from the current source. In the question, this magnetic field is from right to left. The resultant field is the vector sum of the Earth's field and that from the current.

Question	Answer	Mark
3	D	1

Examiner's tip The trace represents an alternating voltage with a peak value of 20 V. The r.m.s. value is 20 V÷√2.

Question	Answer	Mark
4	B	1

Examiner's tip The input power (I_pV_p) is 20 W; if 4 W is wasted, then 16 W is left for output (I_sV_s).

Question	Answer	Mark
5 (a)	The field pattern due to the wire is concentric circles around the wire.	1
	With the current direction shown in the diagram, the direction of the field is anticlockwise (when observed from above).	1
	The solenoid's field is a set of loops, going straight up the middle and looping round the outside.	1
	The direction is from left to right inside the solenoid and from right to left on the outside.	1

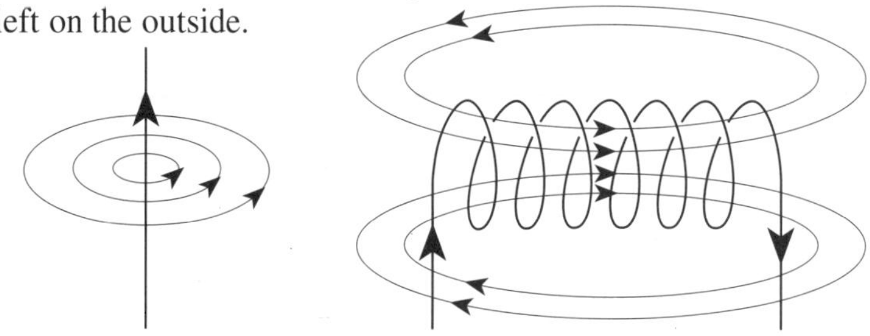

Question	Answer	Mark
(b) (i)	The current in L is up.	**1**

Examiner's tip The fields from the wires at P oppose each other to give a resultant field of zero. The field from K at P is from right to left, so that from L is in the opposite direction which is produced by a current in the same direction as that in K.

Question	Answer	Mark
(ii)	The force is directed towards L.	**1**
(iii)	The force is directed towards K.	**1**
(c)	Along the line from P to Q, the field is parallel to PL in the direction PL.	**1**

Examiner's tip The field along PQ is the vector sum of the fields from K and L. On a plan diagram, these would look like anticlockwise circles. The diagram shows the resultant field at two points along PQ.

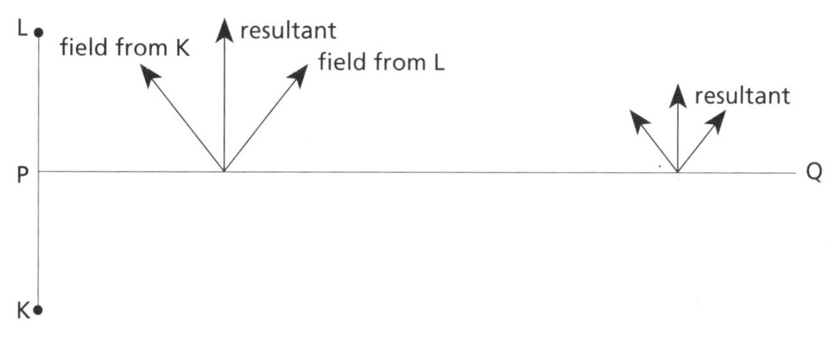

Your graph should show:

	Mark
the field strength increasing from zero at P	**1**
to a maximum value	**1**
then decreasing	**1**

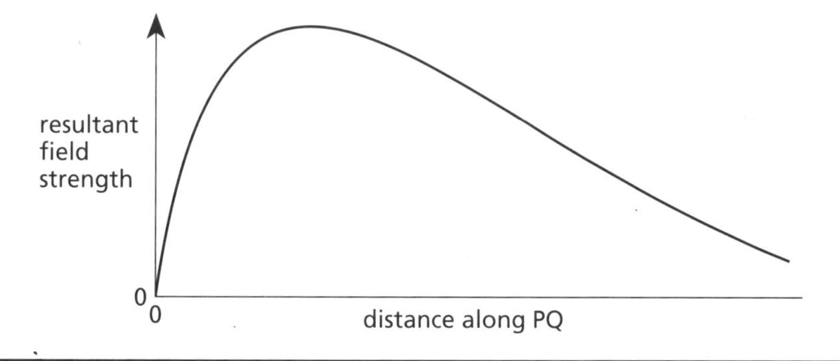

---

6 (a) force = $Bq v$

B = magnetic field strength

q = charge on the ion

v = speed of the ion — **1**

(b) Electric field strength, $E = \dfrac{V}{d}$ — **1**

$$= \frac{6 \times 10^{-4}}{1.4 \times 10^{-3}} = 4.29 \times 10^{-1}\ \text{V m}^{-1}$$ — **1**

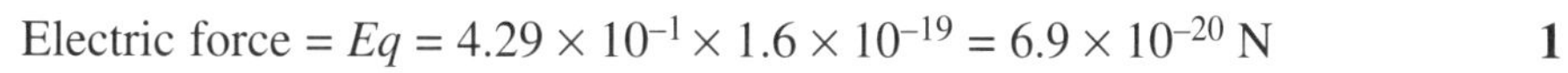

Electric force = $Eq = 4.29 \times 10^{-1} \times 1.6 \times 10^{-19} = 6.9 \times 10^{-20}$ N — **1**

Question		Answer	Mark
(c)	(i)	$Bqv = 6.9 \times 10^{-20}$ N	1
		$v = \dfrac{6.9 \times 10^{-20}}{Bq} = \dfrac{6.9 \times 10^{-20}}{2.0 \times 1.6 \times 10^{-19}} = 2.1 \times 10^{-1}$ m s^{-1}	1
	(ii)	volume per second = speed × cross-sectional area	1
		$= 2.1 \times 10^{-1} \times 1.5 \times 10^{-6} = 3.2 \times 10^{-7}$ m^{3}s^{-1}	1
7 (a)	(i)	Any two from t_2, t_4, t_6, t_8	1
	(ii)	Any two from t_1, t_5, t_9	1
	(iii)	t_3, t_7	1

Examiner's tip The gradient of a displacement-time graph represents the velocity (see Unit 2). The maximum gradient is where the curve crosses the time axis and the gradient is zero at the maximum displacement in each direction.

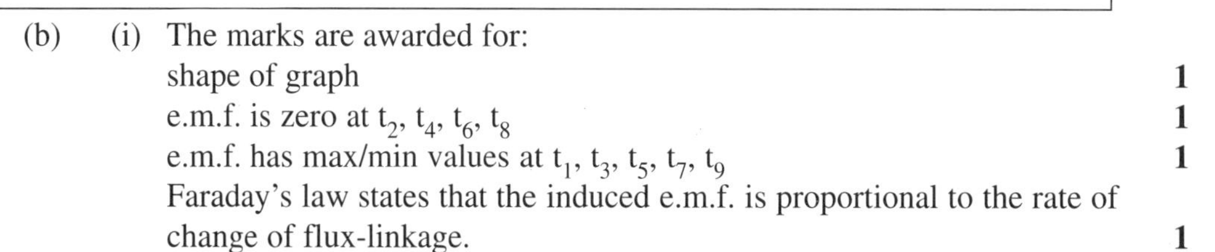

Question		Answer	Mark
(b)	(i)	The marks are awarded for:	
		shape of graph	1
		e.m.f. is zero at t_2, t_4, t_6, t_8	1
		e.m.f. has max/min values at t_1, t_3, t_5, t_7, t_9	1
		Faraday's law states that the induced e.m.f. is proportional to the rate of change of flux-linkage.	1
		This is zero when the speed is zero and a maximum when the speed is a maximum (assuming that the e.m.f. is positive when the magnet is moving upwards).	1

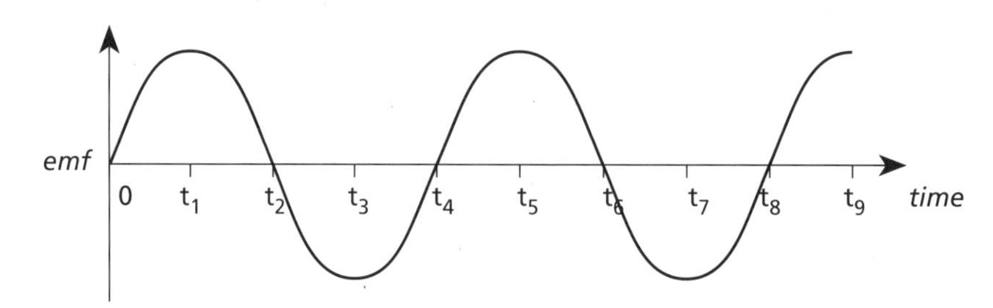

		Lenz's law states that the direction of the induced e.m.f. is such as to oppose the change that causes it.	1
		The change causing the e.m.f. is the movement of the magnet, so the direction of the e.m.f. changes when the direction of motion of the magnet changes.	1

Examiner's tip When the question asks you to use specific laws, you should always quote those laws and relate your answer directly to them.

(c) (i)

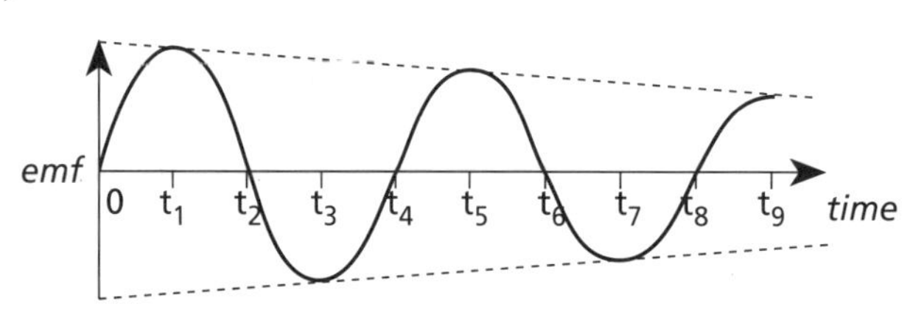

		The graph shows a decreasing amplitude	1
		but no change in frequency.	1

Question	Answer	Mark

Examiner's tip There would be a very slight reduction in the frequency of the oscillation, but with a high resistance this would be negligible.

(ii)	A current now passes in the coil.	1
	The energy comes from the energy of the magnet (which is damped).	1
(iii)	With a low value resistor a large current passes. This causes the coil to have a magnetic field that opposes the motion of the magnet. The effect of the resistive force is to damp the oscillation heavily. This causes the amplitude of oscillation to decrease rapidly and decreases the frequency of the oscillation.	
	Allow one mark for each of the above points up to 4 marks.	4

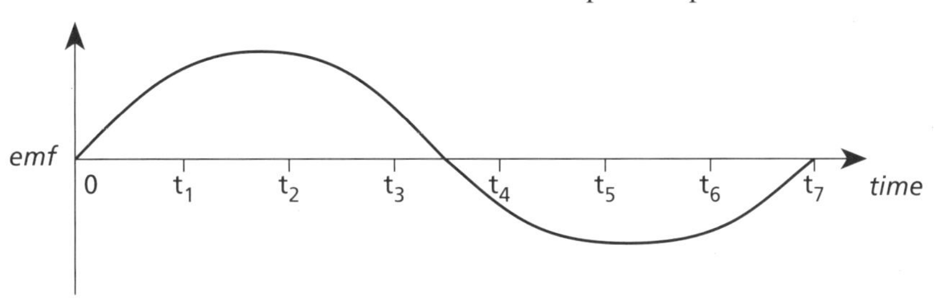

	graph showing reduced frequency	1
	amplitude falling rapidly	1

Examiner's tip Note that the frequency is reduced because the effect of the resistive force is to reduce the resultant force acting on the magnet, so reducing its acceleration (for a given displacement x). This is damped shm. The graph is just one possible way of showing the changes that might occur; the effect of very heavy damping would be to stop the oscillation altogether.

8 (a) (i)	The cable is passing through the magnetic field	1
	at right angles to it.	1

Examiner's tip There would be no e.m.f. generated if the cable were travelling parallel to the magnetic field since there would be no change in flux-linkage. It is also possible to answer this question in terms of the forces on the free electrons in the cable.

(ii)	$E = Blv$	1
	$= 60 \times 10^{-6} \times 20 \times 10^3 \times 7 \times 10^3$	1
	$= 8.4 \times 10^3$ V	1
(b)	The e.m.f. in both cables would be in the same direction	1
	so current could only pass in the same direction in both wires, not 'up' one and 'down' the other.	1

7 THERMAL PHYSICS

Question	Answer	Mark
1	B	1

Question	Answer	Mark

Examiner's tip The energy required for each stage of the process is

energy to melt ice	300 kJ
energy to raise temperature to 100°C	400 kJ
energy to vaporise the water	2000 kJ

The energy required to melt the ice and heat the water to 50°C is 500 kJ, which is 5/27 of the total energy supplied in the one hour time interval.

---

2 **B** **1**

Examiner's tip This is calculated using the formula for centigrade temperature

$$\theta = \frac{100 \times (39.5 - 20.0)}{27.8 - 20.0}$$

---

3 **B** **1**

Examiner's tip The effect of quartering the volume is to quadruple the pressure, while halving the temperature halves it; the overall effect is to double the pressure.

---

4 **A** **1**

Examiner's tip Charge flows in electrical conduction and energy flows in thermal conduction. The rate of energy transfer is analogous to the rate of charge transfer, i.e. the current.

5

Energy gained by the water, $Q = mc\Delta\theta$ (where $\Delta\theta = 21.6°\text{C} - 19.8°\text{C}$) **1**

$= 0.20 \times 4.20 \times 10^3 \times 1.8$ **1**

$= 1.51 \times 10^3$ J **1**

Specific heat capacity of metal, $c = \dfrac{Q}{m\Delta\theta}$ (where $\Delta\theta = 100°\text{C} - 21.6°\text{C}$) **1**

$= \dfrac{1.51 \times 10^3}{0.103 \times 78.4}$ **1**

$= 1.87 \times 10^2$ J kg^{-1} K^{-1} **1**

Examiner's tip The method relies on equating the energy loss by the metal to the energy gain by the water i.e. assuming no other energy losses occur.

Important sources of error are:
energy absorbed by beaker
energy transferred to surroundings while metal is being transferred
energy transfer from water surface Award one mark each for any two. **2**

Ways of reducing these include:
using a lid
transfer the hot metal as rapidly as possible
use a beaker with a low heat capacity For any one: **1**

Question	Answer	Mark

Examiner's tip The way you describe for reducing the error must relate to one of the sources of error that you named.

---

6 (a) As the temperature is constant $p_1V_1 = p_2V_2$ **1**

$$p_2 = \frac{p_1V_1}{V_2} = \frac{400 \times 1000}{250} = 1600 \text{ cm}^3$$ **1**

Examiner's tip You could have worked this out using ratios; the pressure has been multiplied by a factor of 5/8 so the volume has been multiplied by the inverse of this, i.e. 8/5.

(b) The volume stays the same and the pressure increases. **1**
The temperature increases. **1**

Examiner's tip The first two points follow from the graph; the increasing temperature is the only explanation for the pressure having increased.

(c) Considering the change from B to C, as the volume is unchanged,

$$\frac{p_1}{T_1} = \frac{p_2}{T_2}$$ **1**

$$T_2 = \frac{p_2T_1}{p_1} = \frac{500 \times 300}{200} = 750 \text{ K}$$ **1**

Examiner's tip Again, a ratio method could have been used to achieve the same result; alternatively you could have chosen A as the initial state and used

$$\frac{p_1V_1}{T_1} = \frac{p_2V_2}{T_2}.$$

---

7 (a) The warmed water from around the element rises and is replaced by sinking colder water. **1**

(b) The water at the bottom is cold **1**
since the only method of energy transfer to the bottom is by conduction and water is a poor conductor. **1**

(c)

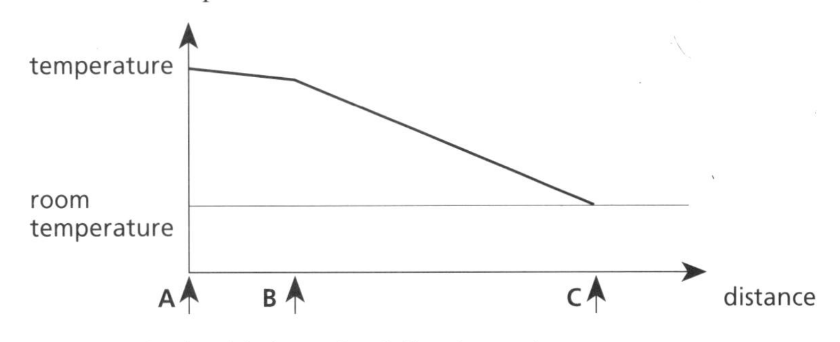

Your graph should show the following points
small rate of temperature change along AB **1**
larger rate along BC **1**
C being at room temperature **1**

Question	Answer	Mark

Examiner's tip Copper is a better conductor than the lagging so a smaller temperature gradient is required to transfer the heat energy at the same rate, as the copper tank and the lagging are in series. C is in thermal contact with the room so it has to be at the same temperature.

---

8 (a) Each 1°C is represented by a distance of 18.0 ÷ 100 = 0.18 cm **1**
The interval from 35°C to 100°C is 65 centigrade degrees **1**
this is represented by a distance of 65 × 0.18 = 11.7 cm **1**

(b) energy transfer = $ml + mc\Delta\theta$ **1**
$= 0.2 \times 3.4 \times 10^5 + 0.2 \times 4200 \times 35$ **1**
$= 6.8 \times 10^4$ J + 2.94×10^4 J = 9.74×10^4 J **1**

---

9 (a) (i) When air is heated it expands. **1**
This causes a decrease in density. **1**
The surrounding, denser air exerts an upward force on it. **1**

(ii) Adiabatic means there is no transfer of thermal energy. **1**

(iii) The first law of thermodynamics states that $\Delta U = \Delta Q + \Delta W$ **1**

In an adiabatic change ΔQ is zero. **1**
Work is done *by* the air on the surrounding atmosphere as it expands, so ΔW is negative **1**
therefore $\Delta U = \Delta W$ is also negative, i.e. the air loses internal energy and cools. **1**

Examiner's tip In the first law of thermodynamics, each term is defined as an increase. It follows that a decrease has a negative value. In this case the expanding gas pushes back the surrounding air doing work on it. This energy transfer is provided by a loss in its internal energy.

(b) (i) The mean temperature is 280 K and the mean pressure is 88 kPa. **1**
using $pV = RT$ (for one mole of gas)
the mean volume, $V = \frac{RT}{p} = \frac{8.3 \times 280}{88 \times 10^3} = 2.64 \times 10^{-2}$ m^3 **1**

density $\rho = \frac{\text{mass}}{\text{volume}} = \frac{0.029}{2.64 \times 10^3} = 1.10$ kg m^{-3} **1**

(ii) pressure due to air = $h\rho g$ **1**
$= 2500 \times 1.1 \times 9.8$ **1**
$= 2.69 \times 10^4$ Pa (or 26.9 kPa) **1**

(iii) Icing on the wings could occur. **1**
The temperature has dropped to freezing point (273 K) and there is water vapour surrounding the aircraft. **1**

Examiner's tip You are not expected to be an aeronautics expert to answer this question; all the information is in the introductory paragraph to (b).

(c) total force = area × pressure difference **1**
$= 0.12 \times 69 \times 10^3$ **1**
$= 8.28 \times 10^3$ N **1**

Question	Answer	Mark
	the tension in each bolt = 8.28×10^3 N ÷ 20 = 4.14×10^2 N	1
10 (a)	Measure the values of the property at two fixed points.	1
	Divide the interval into a number of equal units.	1

Examiner's tip On the centigrade scale the ice and steam points are the fixed points. Any two points can be used that are well-defined and easily reproduced.

Question	Answer	Mark
(b) (i)	The common liquid-in-glass thermometer has:	
	a large bulb to create a measurable change in volume of liquid,	1
	a bulb made of thin glass so that the thermal energy is conducted rapidly to the liquid,	1
	a capillary tube of narrow bore so that the change in volume leads to a relatively large movement of liquid up the stem.	1
(ii)	The liquid-in-glass thermometer is easy to read	1
	but slow to respond/the resistance thermometer can be made to respond more rapidly.	1
	The liquid-in-glass thermometer has a relatively high thermal capacity/ a resistance thermometer can be made with a small thermal capacity.	1
	For accurate readings, the proportion of the liquid-in-glass thermometer that needs to be in the substance varies with temperature.	1
(c) (i)	$\theta = \dfrac{100 \times (940 - 3740)}{210 - 3740}$	1
	$= 79°\text{C}$	1
(ii)	Both properties vary in different non-linear ways between the fixed points.	1

Examiner's tip It is not possible to find a temperature-dependent property that varies linearly between the fixed points; all properties vary in a different non-linear way. The consequence of this is that two thermometers based on different thermometric properties only actually agree at the fixed points.

Question	Answer	Mark
(d) (i)	The absolute scale is based on ideal gases.	1
	It assigns temperatures that do not depend on the instrument used to measure them.	1

Examiner's tip It is important to be able to say 'the temperature is *x*', rather than 'the temperature is *x* if you measure it on this thermometer, but *y* if you measure it on a different one'. The absolute scale is the only scale that allows us to do this.

Question	Answer	Mark
(ii)	$p = \frac{1}{3}\rho < c^2 > = \frac{1}{3}\frac{Nm}{V} < c^2 >$	1

Examiner's tip *N* is the total number of molecules and *m* is the mass of each molecule.

Question	Answer	Mark
	$p = \frac{nRT}{V} = \frac{2}{3} \times \frac{N}{V} \times < E_k >$	1

Examiner's tip The first part of this expression uses $pV = nRT$ to introduce T, the second part substitutes $< E_k >$ for $^1/_2 m < c^2 >$.

Question	Answer	Mark
	So $T = \frac{2}{3} \times \frac{N}{nR} \times < E_k >$ or alternatively $T = \frac{2}{3} \times \frac{N_A}{R} \times < E_k >$	1

Examiner's tip N_A, the number of entities in one mole, is known as Avogadro's number. This expression shows that mean kinetic energy of molecules in an ideal gas is directly proportional to the absolute temperature.

Question	Answer	Mark
(e)	When a substance changes state, e.g. from liquid to vapour, the mean separation of the atoms increases.	1
	This increases their mean potential energy.	1
	But does not affect the mean kinetic energy, which determines the temperature.	1

Examiner's tip It is important to remember that the mean kinetic energy of the atoms and molecules is a measurement of the temperature of a substance.

8 MICROSCOPIC PHYSICS

Question	Answer	Mark
1	**B**	1

Examiner's tip The collisions between molecules are assumed to be elastic; otherwise the gas would lose energy and its temperature would fall.

Question	Answer	Mark
2	**B**	1

Examiner's tip Statement 1 is false; a brittle material would break on the straight-line part of the graph. Statement 2 is correct since the graph is linear for small strains and 3 is correct because the gradient represents stress ÷ strain, which is the definition of Young's modulus.

Question	Answer	Mark
3	**C**	1

Examiner's tip As the cables are in parallel, they share the load so each cable carries a load of 500 kg, which weighs 5000 N. For each cable, $k = 5000 \div 1 \times 10^{-3} = 5 \times 10^{6}\ \mathrm{N\,m^{-1}}$.

Question	Answer	Mark
4	**B**	1

Examiner's tip After one half-life the count rate is 1600 s^{-1}, after two it is 800 s^{-1}, etc. Four half-lives elapse before the count-rate is reduced to 200 s^{-1}, making a time lapse of 1 hour and twenty minutes.

Question	Answer	Mark
5	The nuclide contains 8 protons and 10 neutrons.	1
	The mass of the constituent parts is $8 \times 1.008 + 10 \times 1.009 = 18.154$ u	1
	$E = mc^2 = 0.155 \times 1.660 \times 10^{-27} \times (3 \times 10^8)^2$	1
	$= 2.32 \times 10^{-11}$ J	1

Question	Answer	Mark

Examiner's tip This energy has a mass of 0.155 u; it is important to realise that energy has mass. Einstein's formula can be used to calculate the amount of energy that mass has; it does not state that energy can be changed into mass and vice versa.

---

6 (i) A — **1**

(ii) C — **1**

(iii) B — **1**

Examiner's tip Glass is brittle and it breaks on the linear portion of the stress-strain curve; copper becomes less stiff the more it is stretched while rubber becomes stiffer.

(b) (i) candle-wax — **1**

(ii) nylon — **1**

(iii) diamond — **1**

Examiner's tip An amorphous structure has no regular arrangement; it is like that of a liquid. Some materials appear to be solid but they have no fixed melting point and act like very viscous liquids. Candle-wax and glass are common examples.

(c) (i) Plastic means that the material does not return to the original shape — **1**
once it has been stretched beyond a certain limit. — **1**

(ii) candle-wax — **1**
nylon — **1**

Examiner's tip Nylon is elastic (it does return to its original shape) up to a limit, beyond which it is permanently deformed. Diamond has no plastic phase in its stress–strain curve.

---

7 (a) (i) stress = $F \div A$ — **1**

strain= $(L - L_0) \div L_0$ — **1**

Examiner's tip Strain is the fractional increase in length, i.e. the increase in length divided by the original length.

(ii) From the straight line portion of the graph — **1**
calculate the value of the gradient and take its reciprocal — **1**
(as the graph is plotted with strain along the y-axis).

Examiner's tip Alternatively, choose any pair of values of stress and strain from the straight line part of the graph and calculate stress ÷ strain.

(iii) The work done per unit volume is represented by the area between the line and the strain axis. — **1**

Multiply this by the volume to find the work done. — **1**

Question	Answer	Mark
(b) (i)	Using Pythagoras, the length of the hypotenuse of each triangle is	
	$\sqrt{1000^2 + 100^2} = 1005$ mm	**1**
	L_1 is twice this, i.e. 2010 mm.	**1**
(ii)	$E = \dfrac{Fl}{eA}$	**1**
	$F = \dfrac{FeA}{l} = \dfrac{2.1 \times 10^{11} \times 1 \times 10^{-2} \times 7.8 \times 10^{-7}}{2}$	**1**
	$= 8.19 \times 10^2$ N	**1**
(iii)	$W = 2T \cos \theta$	**1**

Examiner's tip The vertical component of the tension in each wire is $T \cos \theta$; each wire takes half the load.

Question	Answer	Mark
(iv)	$\tan \theta = \dfrac{1000}{100} = 10$	**1**
	$\theta = 84.3°$	**1**

Examiner's tip Because all the sides of the triangle are known, any trigonometric function could be used to calculate the angle.

Question	Answer	Mark
(v)	$W = 2T \cos \theta = 2 \times 8.19 \times 10^2 \times \cos 84.3 = 1.63 \times 10^2$ N	**1**

Question	Answer	Mark
8 (a)	The rate of collisions with the container walls increases.	**1**
	The mean force due to each collision is greater because the mean speed of the molecules is greater.	**1**

Examiner's tip A common error is to state that there are more collisions; this is meaningless without specifying a time. The point is that there are more collisions in any given time interval. Another common misconception is that gas pressure is a result of molecules colliding with other molecules. This is not the case; pressure is a result of the force exerted on the walls of the container.

Question	Answer	Mark
(b) (i)	mass = mass of each atom × number of atoms	
	$= 1.7 \times 10^{-27} \times 1.0 \times 10^7 = 1.7 \times 10^{-20}$ kg	**1**
(ii)	$pV = nRT$ so $p = \dfrac{nRT}{V}$ (where n = number of moles)	**1**
	$n = 1.7 \times 10^{-20} \div 1.0 \times 10^{-3}$ for 1 m^3 of gas	**1**
	$p = \dfrac{1.7 \times 10^{-20} \times 8.3 \times 1.0 \times 10^4}{1.0 \times 10^3} = 1.41 \times 10^{-12}$ Pa	**1**
(iii)	$p = \dfrac{1}{3} nm < c^2 >$	**1**
	$\sqrt{< c^2 >} = \sqrt{\dfrac{3p}{nm}} = \sqrt{\dfrac{3 \times 1.41 \times 10^{-12}}{1.7 \times 10^{-20}}} = 1.6 \times 10^4$ m s^{-1}	**1**

Examiner's tip Note that nm is equal to the mass of 1 m^3 calculated in (b)(i).

Question	Answer	Mark

9 (a) (i)

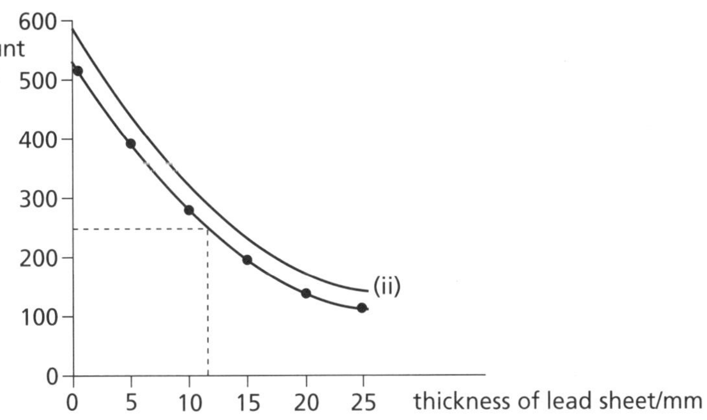

The marks are awarded for:

Correct axes and labels **1**

Points plotted correctly **1**

Smooth curve drawn **1**

The half-value thickness is 11 mm **1**

(ii) The curve is the same shape but every reading is higher. **1**

(b) Four half-lives have elapsed. **1**

The reading would be $^1/_2 \times ^1/_2 \times ^1/_2 \times ^1/_2 \times 520 = 32.5$ counts per minute. **1**

---

10 (a) (i) The decay constant λ is the proportionality constant in the equation rate of decay $\propto$ number of nuclei. **1**

(ii) The half-life, $t_{1/2}$ is the average time taken for the rate of decay of a sample to halve. **1**

Examiner's tip An alternative definition of half-life is the average time for the number of undecayed nuclei to halve.

(b) (i) 27 is the number of protons. **1**

60 is the total number of nucleons (protons and neutrons). **1**

(ii) $\lambda = \dfrac{0.693}{5.26 \times 365 \times 24 \times 60 \times 60} = 4.18 \times 10^{-9}\ s^{-1}$ **1**

Examiner's tip the half-life needs to be in s for the decay constant to be in s^{-1}.

(iii) rate of decay = $\lambda \times$ number of atoms **1**

$= 4.18 \times 10^{-9} \times \dfrac{6.02 \times 10^{23}}{60}$ **1**

$= 4.2 \times 10^{13}\ s^{-1}$ **1**

Examiner's tip This may seem high, but 1 gram is a very large quantity of radioactive material. 1 g of Co-60 has 10^{22} nuclei.

9 DATA ANALYSIS

Question	Answer	Mark

1 (i) The scatter is due to the random nature of radioactive decay. **1**
It could be reduced by taking several readings at each thickness **1**
and calculating an average value, and by counting for longer periods (e.g. 10 minutes). **1**

(ii) The completed table is shown below

x/mm	*N*/min^{-1}	ln(*N*/min^{-1})	*t*/kg m^{-2}
0.00	358	5.88	0
2.00	314	5.75	21.2
4.00	280	5.63	42.4
6.00	248	5.51	63.6
8.00	222	5.40	84.8
10.00	198	5.29	106.0
12.00	178	5.18	127.0
14.00	158	5.06	148.0

The marks are awarded for:
Correct values in the N column (±2 on each reading) **2**
Correct completion of the ln N column **1**
Correct calculations of t **1**

Examiner's tip Note that x must be converted to metres to give t in kg m^{-2}.

(iii) See graph opposite.

The marks are awarded for:
Suitable scales and labelled axes **1**
Plotting of points **3**
Best straight line drawn **1**

Examiner's tip When a grid is provided, choose your scales so that you use it all. If you plot your graph on half of the grid provided, you do not get the first mark. When drawing the best straight line, do not make it go through the first and last points, or as many points as possible. Draw the line with the best-fit gradient, with the points scattered equally above and below the line.

(iv) gradient of graph = $\frac{(5.00 - 5.88)}{158} = 5.57 \times 10^{-3}$ **3**

The marks are awarded for:
gradient calculated as increase in y-value divided by increase in x-value (1)
calculation uses data from a large proportion of the line (1)
correct value (1)

Examiner's tip You should use as much of the line as possible when calculating a gradient from a graph.

$\mu = 5.63 \times 10^{-3}$ m^2 kg^{-1} **1**

Question	Answer	Mark
(v)	The energy is approximately 1.30 MeV.	1
	As μ for the cobalt-60 lies between the values corresponding to energies of 1.20 and 1.40 MeV.	1

5.9
5.8
5.7
5.6
5.5
$\ln(N/\text{min}^{-1})$
5.4
5.3
5.2
5.1
5.0
0 20 40 60 80 100 120 140 160
t/kg m^{-2}

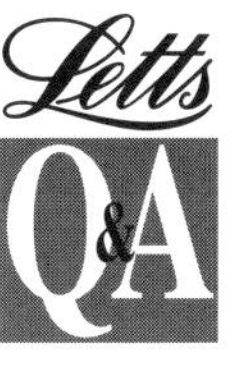